Springer Geography

The Springer Geography series seeks to publish a broad portfolio of scientific books, aiming at researchers, students, and everyone interested in geographical research.

The series includes peer-reviewed monographs, edited volumes, textbooks, and conference proceedings. It covers the major topics in geography and geographical sciences including, but not limited to; Economic Geography, Landscape and Urban Planning, Urban Geography, Physical Geography and Environmental Geography.

Springer Geography—now indexed in Scopus

Luca Salvati

Economic Growth, Regional Disparities, and Urban Sprawl

The Mediterranean Experience

Springer

Luca Salvati
Sapienza University of Rome
Rome, Italy

ISSN 2194-315X ISSN 2194-3168 (electronic)
Springer Geography
ISBN 978-3-031-85798-0 ISBN 978-3-031-85799-7 (eBook)
https://doi.org/10.1007/978-3-031-85799-7

This Springer imprint is published by the registered company Springer Nature Switzerland AG
The registered company address is: Gewerbestrasse 11, 6330 Cham, Switzerland

Contents

Chapter 1
Economic Growth, Landscape, and the City: Toward a Mediterranean 'Sprawl Model'?

Abstract Urban sprawl, a morphological and functional process that reflects settlement dispersion into rural spaces, is a commonly debated issue from different disciplinary perspectives—from spatial planning to environmental science, from applied economics to human geography, from demography to agronomy, from geology to sociology. While implying multiple research dimensions, sprawl is (theoretically and empirically) seen as a sort of 'hegemonic' concept—rather fascinating on the one side, and challenging to investigate and understand on the other side. Earlier studies have tried to define dynamics, implications, and consequences of low-density settlement expansion together on natural environments, social settings, and economic backgrounds. With this perspective in mind, the present contribution provides readers with a detailed picture of the nature of sprawl, focusing on both extrinsic and intrinsic effects on the intimate functioning of local systems. The operational analysis of sprawl focused on representative cases from the Mediterranean region in the old continent, considering narratives and a refined (literature) investigation based on official statistics and traditional—or more innovative—indicators.

Keywords Urbanization · Planning · Sustainable development · Land use · Europe

1.1 Introduction

The study of a so dynamic process such as 'urban sprawl' in contemporary Europe has been carried out extensively implementing a comparative analysis of specific cases dealing with different aspects of this kaleidoscopic phenomenon (Munoz 2003). In this book, we move from land use to population dynamics, from settlement morphology to social segregation, delineating the intrinsic development path of human settlements in the area—assumed to be representative of broader dynamics in the old continent and beyond (Bruegmann 2019). Based on appropriate indicators, the outcomes of earlier studies suggest the consolidation of dominant (urban and metropolitan) trends for the most recent time in Europe (Mouratidis 2019), reflecting

L. Salvati, *Economic Growth, Regional Disparities, and Urban Sprawl*, Springer Geography, https://doi.org/10.1007/978-3-031-85799-7_1

a stratification of semi-dense growth and a more subtle form of sprawl (Cassiers and Kesteloot 2012). Most of the narrative cases—outstandingly explored in a seminal study by Couch et al. (2007)—confirm basic characteristics of sprawl and allow a thorough reflection of the recent metropolitan development in the old continent. Especially in Southern Europe, sprawl was envisaged as a socioeconomic process mixing compact growth and a latent model of city development (Morelli et al. 2014) depending on buildable land scarcity, economic constraints, and contextual factors, together with a still-strong urban ideal typical of the Mediterranean region (Carlucci et al. 2017).

The present chapter illustrates the structure of this book, being composed of distinctive sections answering questions that move from a complete identification of the so-called stylized facts characteristic of sprawl in Mediterranean Europe (Morote and Hernandez 2016). The first part aims at reaching a multidimensional understanding of sprawl, moving from elementary definitions of patterns and processes—and simplified descriptions of background contexts—toward a coherent and, possibly, comprehensive interpretation of this phenomenon, that may consider multiple aspects and issues (Muniz et al. 2003; Millward 2008; Chelleri et al. 2015). The main questions that arise from a literature review of sprawl definitions include the following statements:

(i) What exactly corresponds with the term ‘sprawl’, how is it defined, and how can it be observed, based on the available measurement tools?
(ii) Which are the morphological and functional characteristics of this phenomenon? And, finally,
(iii) which implications does it have on the natural and social environments, and the economic systems?

These questions will initially be addressed with regard to different territorial contexts (Musterd 2005), depending on the availability of primary and secondary information—both theoretical and empirical (Cecchini et al. 2019). Sprawl is, together, related strictly with morphological, environmental, political, social, cultural, and economic characteristics of the territory over which the phenomenon is active (Musterd et al. 2017; Cebrian Abellan and Sanchez Ondono 2019; Nickayin et al. 2020). With this perspective in mind, a comparative scrutiny of the intrinsic and latent characteristics of Mediterranean sprawl may shed light on complex ‘urbanization’ and ‘suburbanization’ dynamics in the old continent (Cazcarro et al. 2012; Orenstein et al. 2014; Oueslati et al. 2015). Following Egidi et al. (2020a),

> to face an increasingly deregulated urban expansion, the quest for interpretive paradigms going beyond traditional urbanism and design approaches is becoming imperative. Integrated planning frameworks should consider together land-use policy, regional economy, and urban sociology, taking a more careful account of the economic structures, productive values, socio-demographic characteristics, and environmental features (e.g., accessibility, land availability, soil quality, and climate) of local territories. (…) A qualitative (visual) investigation of representative case studies may provide the necessary knowledge base for effective assessment and informed policy actions containing dispersed urban expansion and

counteracting the negative effects of sprawl. A context-based analysis of sprawl demonstrates to be particularly appropriate in homogeneous regions—originally developed through compact and continuous urbanization.

The importance of empirical knowledge supporting policy decisions emerges in the fields of economics, statistics, geography, and sociology, which take advantage from information digitization, big data, and advances in national accounts, econometric techniques, and statistical approaches (Pacione 2003; Bianchini et al. 2021). Research at the local scale is becoming a primary objective when integration of the multiple information available to users and practitioners is required (e.g. Biasi et al. 2015). This book illustrates the results of a detailed study demonstrating the intrinsic value of quantitative approaches based on official statistics and territorial indicators when analyzing complex socioeconomic processes (Parcerisas et al. 2012; Pagliarin 2018; Bueno-Suarez and Coq-Huelva 2020). Having a specific focus on sprawl, the work includes different stand-alone contributions connected each other with the investigated topic, the adopted methodologies, and the statistical databases used for a quantitative exploration of this latent phenomenon (Buzar et al. 2007).

In this vein, maps and geographical information tools allow unraveling the spatial complexity of Mediterranean cities (Pardo-Garcia and Merida-Rodriguez 2018) and deepen the operational knowledge of regional geographies focusing on metropolitan regions (Cakir et al. 2008; Cuadrado-Ciuraneta et al. 2017; Carlucci et al. 2018). This will highlight territorial differences within a particular country and between different countries and regions (Catalan et al. 2008). The inherent complexity in any metropolitan region is briefly delineated using a three-way interpretative scheme (Fig. 1.1) based on three dimensions of urban transformation (time, space, and the role of specific sectors as agents of change).

Shaping metropolitan complexity, temporal and spatial factors were further divided into 'fast' or 'slow' variables, shifting (rapidly or slowly) over a given time horizon, as well as 'broad range' and 'narrow range' variables, respectively creating homogeneous or more heterogeneous spaces because of the geographical scale they act. The book also provides recommendations to policy makers and scholars of

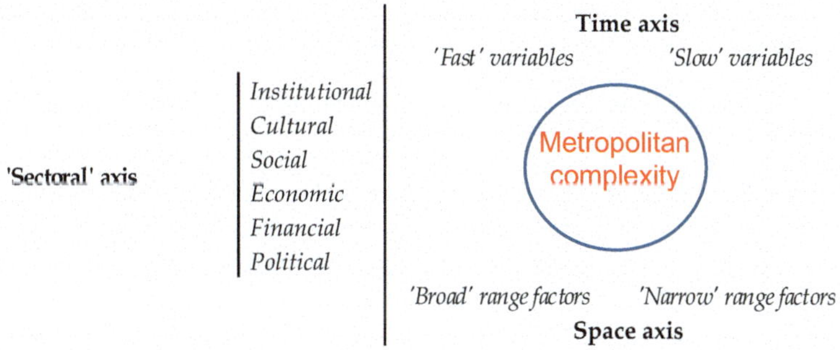

Fig. 1.1 An interpretative scheme of 'metropolitan complexity' based on three dimensions of urban transformation (time, space, and the role of specific sectors as agents of change); source: author's elaboration of literature evidence presented in this chapter

advanced countries that may stimulate a deeper understanding of the functioning of local systems (sensu Paul and Tonts 2005; Chorianopoulos et al. 2010; Pares et al. 2013).

Moving from theory to practice, the subsequent part of the work represents a first step toward the objective of a redefinition and a full rethinking of the hegemonic notion of 'sprawl' (Polyzos et al. 2013; Christodoulou and Oikonomou 2018; Perrin et al. 2018). With regard to the last issue, the concept of an individual form and structure of Mediterranean sprawl will be introduced under the assumption that this phenomenon varies within the same continent, and even the same geographical region (Recanatesi et al. 2016; Pozoukidou and Ntriankos 2017; Ciommi et al. 2018, 2019). Readers will be provided with a complete description of the most characteristic (local) development trajectories in the Mediterranean region, by introducing a classification of urban trends and metropolitan functions, starting from some seminal studies, such as Clark (1951). Consequently, using data released from national and supra-national organizations, the impact of Mediterranean sprawl on local societies and environments will be delineated and quantified (Colaninno 2012). A preliminary, and barely qualitative, definition of 'sprawl' may thus stem from the following considerations, in line with the interpretative framework exemplified in Egidi et al. (2020a), as a mostly unplanned (or informal).

> form of urban growth in which large- and medium-size cities expand into rural areas with low-density discontinuous settlements, creating mixed peri-urban landscapes. As a result of these changes, landscape become messy and confused, characterized by fragmented and unstable uses of land, with a series of negative effects for the environment (e.g. the loss of natural values and the negative impact of commuting) and the social sphere (e.g. the loss of identity of the rural population, class segregation, social homogenization, and economic polarization).

Based on such a rationale, the first part of our book is therefore a step toward the objective of defining sprawl in Southern European cities. The nature and intrinsic dynamics of this process are clearly the outcome of geo-economic and sociocultural characteristics of the territory and its settlements (Power 2001; Colantoni et al. 2017; Robert et al. 2019). Following the seminal work by Galster et al. (2001), urban sprawl

> has variously been used to refer to: patterns of urban development; processes of extending the reach of urbanized areas; causes of particular practices of land use; and the consequences of those practices. (...) One of the advantages claimed for this definition is that it suggests and can accommodate different types of sprawl. Furthermore, it permits sprawl to be considered as a process and not merely a pattern of urbanization.

Consequently, in order to understand the true nature of Mediterranean sprawl, the book provides a historical, geo-economic, and sociocultural description of Mediterranean landscapes and societies (Rontos et al. 2016), assuming landscape traits and demographic dynamics (Fig. 1.2) as the main components of the multivariate and kaleidoscopic profile of urbanization in advanced economies (Coq-Huelva and Asian-Chaves 2019). In this perspective, Egidi et al. (2020a) argue that

> a more comprehensive interpretation of multiple sprawl processes was articulated in different research dimensions [37], providing an operational definition of the phenomenon as a pattern

> of land-use in an urbanized area that exhibits low levels of some combination of distinct dimensions, that include (i) density, (ii) continuity, (iii) concentration, (iv) clustering, (v) centrality, (vi) nuclearity, (vii) mixed land-use, and (viii) proximity. Although much of these dimensions can be associated with population trends, socio-demographic transitions, and economic dynamics, some of these attributes are specifically linked to different typological and socio-spatial settlements.

Based on this conceptual definition, the kaleidoscope of morphologies and functions reveals the inherent complexity in any metropolitan region of the area, and requires a multidisciplinary interpretation of patterns and processes of urbanization, mixing quantitative and qualitative approaches (Salvati et al. 2016). However, in line with the seminal study by Couch et al. (2007), some operational definitions tried

> to emphasize the idea of urban sprawl being a type of urban form or a pattern of urbanization, rather than a process of urban change. However, in our view, the latter may be a more useful perspective, since it is the process of sprawling that leads to undesirable side effects and it is in the process of sprawling that policy must intervene. A feature of our discussion is therefore that it concentrates on urban sprawl, not as a pattern of urbanization, as is more usual in the literature, but rather as a process of urban change.

Supported by qualitative and quantitative analysis, the academic discussion on sprawl—as reflected in this book—has basically focused on the comprehension of the

Fig. 1.2 A kaleidoscope of indirect signs of sprawl in the Mediterranean region (from upper left to lower right, clockwise: morphological expansion of residential settlements on greenfields and brownfields in Rome; fringe settlements and forest fires in Athens; the oneiric linkage between desertification and urbanization in Barcelona, courtesy of the Water Regional Agency of Catalonia; and the peri-urban setting of relict horticulture in Naples). *Sources* Author's photographic archive

Mediterranean urban ideal and how cities of this region are shaped by certain behaviors and practices of local communities (De Muro et al. 2011; Rubiera-Morollon et al. 2016; Salvati et al. 2019). Finally, at the end of this analysis, the main characteristics of the metropolitan development of Northern Mediterranean and Anglo-American societies will be resumed and confronted, with the aim at delineating the 'intrinsic' nature of Southern European sprawl styles, if any (Di Feliciantonio et al. 2018; Salem et al. 2020; Rodrigo-Comino et al. 2021).

The final part of the book delineates a comprehensive picture on the present and future state of discontinuous urbanization and exurban development in the Northern Mediterranean basin (Diaz-Pacheco and Garcia-Palomares 2014; Sallustio et al. 2018; Rubiera-Morollon and Garrido-Yserte 2020). This section consolidates the most recent theoretical and empirical advancements in geo-economic literature (Doygun 2009; Salvati 2014; Rontos et al. 2022): methodologies and results achieved by different scholars when assessing sprawl were critically discussed (e.g. Durà-Guimera 2003). The analysis has been mainly based on the results of empirical studies carried out on secondary data—basically official sources (Salvati and Carlucci 2016).

In the light of defining the nature of sprawl in the old continent, the reasons that have brought this study to the choice of Mediterranean urbanities as an appropriate background context for a qualitative and quantitative analysis of past, present, and future urbanization patterns and processes will be finally explained and discussed (Di Feliciantonio and Salvati 2015). Additionally, assuming urbanization as a powerful expression of the cultural traits of any human society, a comprehensive picture of the 'dominant' urban culture in the Mediterranean basin will be delineated (Richardson and Chang-Hee 2004). Being challenging to identify and interpret coherently, social dynamics have experienced many breakpoints (Fig. 1.3), based on intense cultural, economic, and demographic flows from different continents (Domene and Sauri 2006; Darques 2015; Rejeb Bouzgarrou et al. 2019).

Urban sprawl should be understood as a particularly complex and nonlinear phenomenon, which depends on multiple economic, social, and ecological factors (Egidi et al. 2020a). Scientific literature clearly shows the difficulty of correctly defining the phenomenon, which assumes strong territorial declinations and specificities in the various regions and territorial contexts of the world. Identifying the triggering drivers, in a general or regional perspective, is also a demanding research task (Gross and Ouyang 2021). Avoiding inappropriate generalizations, this book focuses on dispersed urbanization in Europe, with a specific focus on the Mediterranean region—understood as paradigmatic, but at the same time representative, of more general trends on a continental scale (Ji et al. 2020). Figure 1.3, therefore, remains faithful to this theoretical assumption, which informs and guides the research and documentation path proposed in this book. Urban sprawl is read, in its complexity, through a DPSIR (Driving forces-Pressures-State-Impacts-Responses) interpretative scheme, namely a well-known framework for environmental reporting at national and community scale (Shaw et al. 2020).

DPSIR scheme revealed to be particularly useful when defining the co-presence of potentially important factors underlying specific environmental and socioeconomic

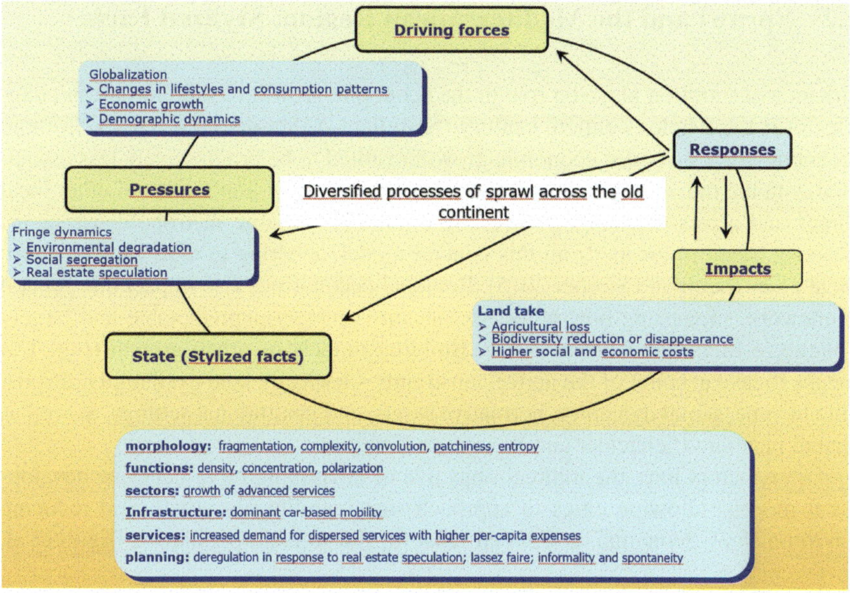

Fig. 1.3 A stylized picture illustrating diversified patterns and processes of sprawl in the old continent and the most pertinent underlying factors; source: author's elaboration of literature evidence presented in this chapter

processes worthy of attention (Di Clemente et al. 2021). In particular, the scheme focuses on a permanent monitoring of sprawl, delineating some original dimensions of study, also based on quantitative indicators (Salvati et al. 2016). The dimensions that attract the greatest attention here are settlement forms, the economic functions of cities and the metropolitan regions at large, the production base and the subdivision into economic activity sectors, the availability of infrastructures, and the structuring of networks that guarantees accessibility from the outside and mobility to residents (Colantoni et al. 2017). The ensemble of services provided to the population, as well as the set of planning responses to the issues that the expansion of dispersed settlements sets up from time to time, possibly add as new challenges to regional sustainability (Egidi et al. 2020b).

In this perspective, the 'urban culture-city configurations' nexus in the post-war Mediterranean European cities was specifically analyzed. Literature review and empirical findings reveal that Southern Europe results as heterogeneous in some aspects (i.e. spatial forms, economic functions), but similar in others, including social dynamics and behaviors (Salvati et al. 2018; Salvia et al. 2022; Schwaab 2022). Rejecting a unique, generalized model for the interpretation of local development processes in Mediterranean cities (Sauri 2003; Domene et al. 2005; Sassen 2010), however, does not imply that there are no common characteristics that the main urban nodes of this region may share.

1.2 Sprawl and the Mediterranean Region: Stylized Facts

Urban concentration played a role in the economic growth of Mediterranean countries over the whole twentieth century (Seifollahi-Aghmiuni et al. 2022). In more recent decades, however, economic growth resulted to be progressively less associated with the rate of population growth, suggesting rising importance of other local forces and actors of change (Duvernoy et al. 2018). Large metropolitan regions, however, seem to escape from this general model, adhering to individual development paths (Scott and Storper 2015). Regional peculiarities also impact this general framework, suggesting how cities' expansion is mostly unpredictable and largely volatile (e Silva and Correia 2023). Building models of urban growth (Fig. 1.4) means to take account of the active constraints—basically land availability, disposable income, social dynamics, normative assets, and institutional settings, as well as spatial planning (Schneider and Woodcock 2008).

These factors have the indirect objective of envisaging more cohesive development models, allowing cities to approach sustainability objectives, and reducing environmental, economic, and social risks for the resident population (Egidi et al.

Fig. 1.4 Dominant spatial direction of radio-centric expansion of residential settlements in metropolitan Athens' districts between 1945 and 2020; source: author's elaboration of official statistics from census of buildings

2020b). According to this rationale, sprawl is a multifaceted phenomenon that needs to be quantified with distinctive measures over sufficiently long observation times (Salvati et al. 2013). Landscape change is recognized, in this perspective, to be a relevant issue when assessing the impact of urban growth and sprawl on fringe land (El Menchawy et al. 2011; Delfanti et al. 2016; Escudero Gomez et al. 2019).

In the kaleidoscopic panorama of the old continent, Mediterranean cities represent elements that break with the stereotypes of European urbanities (European Environment Agency 2006). Originally developed following purely mono-centric structures (Salvati 2019), which grew radio-centrically with the progressive addition of satellite urban areas, the large Mediterranean cities have more recently oriented their growth path toward dispersed settlement models (Feleki et al. 2018). This resulted in a high rate of land consumption (Feng et al. 2009), in contrast with the 'judicious compactness' that has characterized the tumultuous development of such cities at least in the first three decades after World War II (Salvati and Gargiulo Morelli 2014).

With these perspectives in mind, the book describes—using narratives, photographs, qualitative-quantitative descriptions of landscapes, maps, and indicators (Fig. 1.5)—the transition from a compact model—'judiciously mono-centric' and 'land-saving', to contemporary sprawl (Feng et al. 2019). The environmental, social, and economic implications of this phenomenon, far from being a unique characteristic of the Mediterranean region, have been extensively discussed (Ferrara et al. 2016), and the planning and policy implications addressed with respect to the individual ecological problems faced from time to time (Ferrara et al. 2017).

In this vein, Chap. 2 discusses the intrinsic mosaic made up of specialized and cooperative networks forming the largest Mediterranean cities, displaying a relatively high level of development, territorial cohesion, and competition on a global scale (Fuladlu et al. 2021). The polycentric model, formed by specialized production clusters governed by high-level external economies therefore appears, at least from an economic point of view, to be a correct ground for regional planning (Garcia-Lopez 2019). Polycentric structures represent a decisive reference in policy and planning choices at all spatial scales (Garcia-Lopez and Muniz 2010). From the economic perspective, urban expansion in the Mediterranean region is regarded as an intriguing (and possibly intricate) issue, given its unpredictable nature (Garcia-Coll and Lopez-Villanueva 2018). Urbanization—both compact and dispersed—has occurred in different ways across cities, highlighting peculiar relationships between urbanization, land use, and economic structures (Garcia-Ayllon 2013).

More recently, settlement dispersion advanced rapidly in Mediterranean Europe—irrespective of city's size and morphology—with urbanization rates growing faster than total population (e.g. Galster et al. 2001), as documented in Chap. 3. Despite sharing common drivers of change (related to population dynamics, sociospatial structures, and deregulated settlement expansion), metropolitan growth has adapted to the local economic, cultural, sociodemographic, and environmental context (Garcia-Nieto et al. 2018). Understanding how place-specific factors influence urbanization processes in Mediterranean contexts provides suggestions for appropriate policies of settlement containment and land-use management, especially targeting the uneven expansion of large cities (Giannakourou 2005).

Fig. 1.5 Vacant, degraded land separating compact (but spatially discontinuous) buildings from low-density, sprawled settlements in Oeiras, Eastern Lisbon; source: author's photographic archive

From a social perspective, urbanization is considered as the one of the most powerful expressions of the cultural traits of any human society (Gospodini 2009). However, a comprehensive picture of the urban culture in the Mediterranean area has been rather difficult to construct, as it has experienced conflicts and intense flows from different trade areas in the world (Hennig et al. 2015). In this vein, Chap. 4 proposes a review of the 'urban culture–city configuration' nexus in post-war 'Southern' cities. This work focuses on the particular dynamics of Mediterranean societies together with the informal economic system, which has played a crucial role in the long-term development of entire regions (Gavalas et al. 2014; Sharaf et al. 2018; Soulard et al. 2018).

This chapter refines and completes the focus on social dynamics in 'Southern' cities by discussing the inherent trends in the working class (Fig. 1.6), a population segment that has been rarely associated with a single mode of land and housing allocation (Terzi and Bolen 2009). While representing a relatively large portion of a given society (Habibi and Asadi 2011), the latent evolution of such class was discussed in light of the implications for economic growth and the cycle from urbanization to suburbanization in Mediterranean economies (Tsilimigkas et al. 2018). The location and tenure patterns of the working class are constrained by a series of latent forces, which result in the emergence of a variety of housing conditions (Venanzoni et al. 2017; Varela-Candamio et al. 2019; Halbac-Cotoara-Zamfir et al.

Fig. 1.6 One of the most attended, spontaneous manifestations in Platia Syndagmatos, downtown Athens, during the 2015 summer coinciding with the 'οχι' [No] referendum about 'Troika' memorandum in Greece; source: authors' photographic archive

2021). Finding a theory that may successfully explain working-class land allocation patterns is complicated, as this is influenced by a kaleidoscope of economic forces representative of urban and rural societies, and the specific system of production, distribution, and consumption (Turok and Mykhnenko 2007; Townsend 2009; Vidal et al. 2011).

1.3 Local Development, Sprawl, and the Mediterranean Legacy

Moving from a sort of a structural crisis—characteristic of Southern Europe since decades—Chap. 5 reflects on a vast set of theoretical approaches and empirical examples contributing to shift toward a resilience discourse in urban affairs. Focusing on both morphological and functional issues (Gielen et al. 2018), these frameworks may provide an appropriate vision for interpreting metropolitan complexity in an upcoming 'metropolitan' world (Tsapala et al. 2023). Within this context, resilience of metropolitan regions can be understood as the ability to adapt to economic, technological, and political changes, affecting evolutionary dynamics and trajectories pursued by regional economies (Gonzales 2011; Storper and Scott 2016; Tomao et al. 2021).

This contribution also investigates territorial dynamics in light of the demographic, economic, and environmental processes that underlie urban–rural configurations, suggesting that polycentrism—understood as a morphological and functional feature of metropolitan structures—is not sufficient to mobilize actions toward sustainable development, social cohesion and environmental conservation of target ecosystems (Tsilimigkas et al. 2016; Halbac-Cotoara-Zamfir et al. 2022; Vinci et al. 2022). The evolution of environmental processes over time and space was finally discussed, assuming the intrinsic role of socioeconomic dynamics and the possible relationship with regional development paths associated with settlement polycentrism and spatially balanced development (Ward 2010; Grout et al. 2016; Vinci et al. 2023).

Trends regarding population growth and urbanization rates in the Mediterranean region were specifically investigated in this perspective. 'Coastalization', one of the most frequent development patterns in the region (Fig. 1.7), will be discussed widely, taking advantage from multiple data sources and qualitative and quantitative information (Hepcan et al. 2013; Wolff et al. 2020; Yilmaz and Terzi 2021). Some words will be also spent in demonstrating the connection between over-urbanization and growing vulnerability of cities toward natural disasters (Hof et al. 2013; Zambon et al. 2018, 2019).

An empirical analysis of the observed consequences of sprawl, and more specifically of 'coastalization', will document how these complex relationships require a comprehensive analysis from a joint perspective of environmental sustainability and spatial planning. This analysis may also suggest suitable strategies coping with land degradation (Zwiers et al. 2016; Zambon et al. 2017; Ferrara et al. 2020). In a time of crisis, an increasing debate has been promoted on the sustainable management of peri-urban land, including agricultural and industrial activities located across fringe and on the related urban–rural relations (Ismael 2021). In a perspective of sustainable development (Kaika 2003), urban peripheries can be seen as multifunctional locations that experience diversified economies. These have frequently promoted agricultural land use intended as a physical buffer to dispersed settlement expansion (Kaniewski et al. 2013), in turn triggering local development paths grounded on natural amenities.

In this perspective, a reflection on peri-urban landscapes under economic crisis was proposed integrating different disciplinary visions (Karadimitriou et al. 2021). Non-urban, non-rural areas are the objective of this approach, going beyond the aesthetic-visual rationale typical of landscape studies, and reflecting a spatial economic perspective (Kasanko et al. 2006). When addressing territorial complexity, a strong integration between quantitative methodologies and qualitative approaches is increasingly required, especially in a multidimensional analysis of metropolitan growth and change (e.g. Kosmas et al. 2015). Evaluating latent landscapes under crisis requires an in-depth understanding of historical, cultural, and political implications of economic recession (Keddie and Tonkiss 2010), suggesting action plans that can

Fig. 1.7 Toward a 'Mediterranean dream'? Staying alone in the summer of peri-urban Schinias (North-Eastern Attica), in between the sea and the pinewoods, one kilometer far away from the Olympic rowing center; source: author's photographic archive

trigger virtuous processes of sustainable development and revitalize degraded environments in a time of reduced public budget and increasing social inequalities in metropolitan regions (Kemeny and Storper 2012).

1.4 From Structure to Functions: Unraveling the 'Intrinsic Manifestation' of Sprawl

Economic growth may promote the conservation of suburban land and the inherent recovery of degraded environments along the interface between settlements and natural habitats (Zambon and Salvati 2019). The potential for economic growth and social wealth also depends on the quality of the environment, the ecological carrying capacity, and the appropriateness of spatial planning targeting the maintenance of land capital (Kosmas et al. 2016). Landscape protection and recovery in suburban districts was thus aimed at rebalancing urban and rural areas into a new 'metropolitan continuum' (Koprowska et al. 2020). By highlighting the strict connection between sprawl and the intrinsic preservation of rural landscapes in advanced economies (Ward 2010), a brief reflection on the relationship between fringe landscapes and

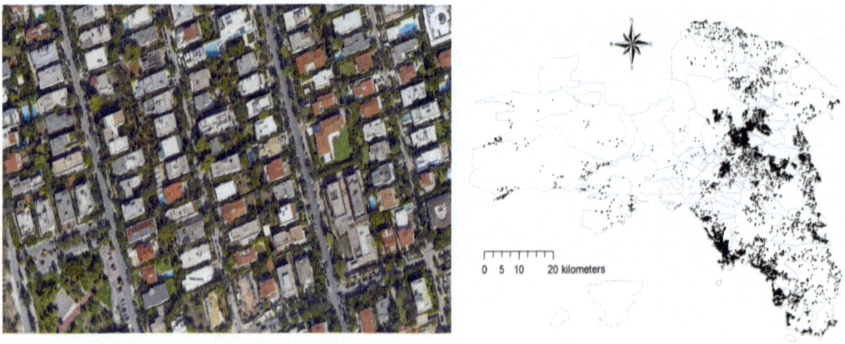

Fig. 1.8 Spatial distribution of residential swimming pools in metropolitan Athens as an indirect sign of sprawl and social segregation; source: author's elaboration (left) and visual interpretation (right) of Google Earth imagery

suburban farming was finally introduced, discussing the possible implications for rural development (Barbero-Sierra et al. 2013).

A framework quantifying and planning the multiple relationships between agro-forest systems and human settlements in Mediterranean Europe demonstrated to be partly ineffective when it concentrates on individual—even if pivotal—policy targets (Kourliouros 1997; Lagarias and Sayas 2019; Lagarias and Prastacos 2020). Eco-sustainable planning, integrated with the multifaceted economic dimensions and social dynamics characteristic of Mediterranean suburban districts seems to be an appropriate way to secure fringe land sustainability (Lagarias and Stratigea 2023). A permanent assessment of such factors allows for the implementation of different development scenarios (Fig. 1.8). This discussion finally contributes to systemic and multiscale approaches informing environmental policies, with the aim at achieving an integrated management of suburban landscapes (Marull et al. 2009).

Urbanization has been one of the most debated topics in both scientific and practitioners' fields (Kovacs et al. 2019). The book concludes by focusing on recent metropolitan development in the Mediterranean region. As a matter of fact, contemporary cities are reorganizing over broader spaces, dispersing over vast territories and long distances, and mixing with the countryside (Kroll and Kabisch 2012). Metropolises are occupying empty or rural spaces at very fast rates, with a mechanism reflecting the increased decoupling from central place locations—in line with a process of globalization involving advanced economies and emerging countries (Le Goix 2005). The distance from inner cities is thus losing its relevance as an ordering principle of settlements along metropolitan gradients (Zambon and Salvati 2019).

Accounting for multifaceted development paths that span from compactness to diversified degrees of settlement dispersion, contemporary cities have historically expanded under different economic conditions (Lee 2011). From a spatial perspective, the configuration of a number of metropolitan regions in the old continent reflects a latent transition from hierarchical and polarized settlement structures to more heterogeneous and fragmented models (Leontidou 2009). A mosaic of specialized and cooperative functional networks, with a greater level of technology diffusion, territorial cohesion, and competitive capacity, consolidated as a result of exurban development (Liu and Meng 2020). The polycentric model, formed by specialized production clusters governed by higher-level (external) economies appears, in this vision, an appropriate way to plan heterogeneous and socially divided territories (Lindstrom and Bartling 2003). The polycentric model thus became a conceptual reference in any policy decision toward sustainability at all the governance scales, from national to local.

1.5 Concluding Remarks

This study concerns the complex process of urban sprawl and its determinants in the context of Mediterranean Europe. The focus on the specificity of the analyzed process in this part of the old continent gives an added value to a broader empirical literature on urbanization and economic growth in advanced economies (Pradham 2017). Within the current metropolitan context—characterized by a marked fragmentation of population, settlements, and activities—qualitative and quantitative approaches that identify new forms of urbanization are increasingly required to delineate a more sustainable relationship between forms and functions (Longhi and Musolesi 2007). Moving toward a land-use approach focusing on settlement structure, the modalities of settlement expansion were demonstrated to be basically dependent on the interplay among socioeconomic aspects, territorial constraints, and planning regulations (Marti et al. 2017).

The present study finally stimulates further studies in the processes of metropolization, suburbanization, and their consequences for the spatial differentiation of socioeconomic development. By reconnecting a morphological analysis of mixed (urban–rural) landscape mosaics with a functional characterization of entire regions (Evers et al. 2024), this book contributes to urban studies, rethinking the importance of spatial planning in the intrinsic regulation of exurban development (Mehriar et al. 2020), especially in socioeconomic contexts where informal settlements had been recognized as the engine of growth for decades (Masini et al. 2019). In this ambit, sustainable development strategies oriented toward settlement containment and planning for resilient cities should interface more effectively with local contexts, benefiting from place-based actions and local practices.

References

Barbero-Sierra C, Marques MJ, Ruíz-Pérez M (2013) The case of urban sprawl in Spain as an active and irreversible driving force for desertification. J Arid Environ 90:95–102

Bianchini L, Marucci A, Sateriano A, Di Stefano V, Alemanno R, Colantoni A (2021) Urbanization and long-term forest dynamics in a metropolitan region of Southern Europe (1936–2018). Sustainability 13(21):12164

Biasi R, Brunori E, Smiraglia D, Salvati L (2015) Linking traditional tree-crop landscapes and agro-biodiversity in Central Italy using a database of typical and traditional products: a multiple risk assessment through a data mining analysis. Biodivers Conserv 24(12):3009–3031

Bruegmann R (2019) Sprawl: a compact history. University of Chicago Press, Chicago

Bueno-Suárez C, Coq-Huelva D (2020) Sustaining what is unsustainable: a review of urban sprawl and urban socio-environmental policies in North America and Western Europe. Sustainability 12(11):4445

Buzar S, Ogden P, Hall R, Haase A, Kabisch S, Steinfiihrer A (2007) Splintering urban populations: emergent landscapes of reurbanisation in four European cities. Urban Stud 44(4):651–677

Cakir G, Un C, Baskent EZ, Kose S, Sivrikaya F, Keles S (2008) Evaluating urbanization, fragmentation and land use/cover change pattern in Istambul city, Turkey from 1971 to 2002. Land Degrad Dev 19:663–675

Carlucci M, Grigoriadis E, Rontos K, Salvati L (2017) Revisiting a hegemonic concept: long-term 'Mediterranean urbanization' in between city re-polarization and metropolitan decline. Appl Spat Anal Policy 10:347–362

Carlucci M, Chelli FM, Salvati L (2018) Toward a new cycle: Short-term population dynamics, gentrification, and re-urbanization of Milan (Italy). Sustainability 10(9):3014

Cassiers T, Kesteloot C (2012) Socio-spatial inequalities and social cohesion in European cities. Urban Stud 49(9):1909–1924

Catalán B, Saurí D, Serra P (2008) Urban sprawl in the Mediterranean? Patterns of growth and change in the Barcelona Metropolitan region 1993–2000. Landsc Urban Plan 85(3–4):174–184

Cazcarro I, Duarte R, Sánchez-Chóliz J (2012) Economic growth and the evolution of water consumption in Spain: a structural decomposition analysis. Ecol Econ 96:51–61

Cebrián Abellán F, Sánchez Ondoño I (2019) Urban sprawl in inner medium-sized cities: the behaviour in some Spanish cases since the beginning of the 21st century. Urban Sci 3(1):10

Cecchini M, Zambon I, Pontrandolfi A, Turco R, Colantoni A, Mavrakis A, Salvati L (2019) Urban sprawl and the 'olive' landscape: sustainable land management for 'crisis' cities. GeoJournal 84(1):237–255

Chelleri L, Schuetze T, Salvati L (2015) Integrating resilience with urban sustainability in neglected neighborhoods: challenges and opportunities of transitioning to decentralized water management in Mexico City. Habitat Int 48:122–130

Chorianopoulos I, Pagonis T, Koukoulas S, Drymoniti S (2010) Planning, competitiveness and sprawl in the Mediterranean city: the case of Athens. Cities 27:249–259

Christodoulou C, Oikonomou M (2018) Urban sprawl typologies in medium-sized Greek cities. A qualitative morphological analysis in the peri-urban landscape. Urban Morphol 16:18

Ciommi M, Chelli FM, Carlucci M, Salvati L (2018) Urban growth and demographic dynamics in southern Europe: toward a new statistical approach to regional science. Sustainability 10(8):2765

Ciommi M, Chelli FM, Salvati L (2019) Integrating parametric and non-parametric multivariate analysis of urban growth and commuting patterns in a European metropolitan area. Qual Quant 53(2):957–979

Clark C (1951) Urban population density. J Roy Stat Soc 114:490–496

Colaninno N (2012) Sustainable territorial development and urban growth: a critical interaction: the Spanish Mediterranean Coast, and Catalonia, during the last two decades. ACE Arch City Environ 7(20):149–172

Colantoni A, Grigoriadis E, Sateriano A, Sarantakou E, Salvati L (2017) Back to Von Thunen: a Southern European perspective on mono-centric urban growth, economic structure and non-urban land decline. Int Plan Stud 22(3):173–188

Coq-Huelva D, Asián-Chaves R (2019) Urban sprawl and sustainable urban policies. A review of the cases of Lima, Mexico City and Santiago de Chile. Sustainability 11(20):5835

Couch C, Petschel-Held G, Leontidou L (2007) Urban sprawl in Europe: landscapes, land-use change and policy. Blackwell, London

Cuadrado-Ciuraneta S, Durà-Guimerà A, Salvati L (2017) Not only tourism: Unravelling suburbanization, second-home expansion and "rural" sprawl in Catalonia, Spain. Urban Geogr 38(1):66–89

Darques R (2015) Mediterranean cities under fire. A critical approach to the wildland–urban interface. Appl Geogr 59:10–21

De Muro P, Monni S, Tridico P (2011) Knowledge-based economy and social exclusion: shadow and light in the Roman socio-economic model. Int J Urban Reg Res 35(6):1212–1238

Delfanti L, Colantoni A, Recanatesi F, Bencardino M, Sateriano A, Zambon I, Salvati L (2016) Solar plants, environmental degradation and local socioeconomic contexts: a case study in a Mediterranean country. Environ Impact Assess Rev 61:88–93

Di Feliciantonio C, Salvati L (2015) 'Southern' alternatives of urban diffusion: investigating settlement characteristics and socio-economic patterns in three Mediterranean regions. Tijdschr Econ Soc Geogr 106(4):453–470

Di Clemente R, Strano E, Batty M (2021) Urbanization and economic complexity. Sci Rep 11(1):3952

Di Feliciantonio C, Salvati L, Sarantakou E, Rontos K (2018) Class diversification, economic growth and urban sprawl: Evidences from a pre-crisis European city. Qual Quant 52(4):1501–1522

Díaz-Pacheco J, García-Palomares JC (2014) Urban sprawl in the Mediterranean urban regions in Europe and the crisis effect on the urban land development: Madrid as study case. Urban Stud Res

Domene E, Sauri D (2006) Urbanization and water consumption: influential factors in the metropolitan region of Barcelona. Urban Stud 43(9):1605–1623

Domene E, Sauri D, Parés M (2005) Urbanization and sustainable resource use: the case of garden watering in the metropolitan region of Barcelona. Urban Geogr 26(6):520–535

Doygun H (2009) Effects of urban sprawl on agricultural land: a case study of Kahramanmaraş, Turkey. Environ Monit Assess 158:471–478

Durà-Guimera A (2003) Population deconcentration and social restructuring in Barcelona, a European Mediterranean city. Cities 20(6):387–394

Duvernoy I, Zambon I, Sateriano A, Salvati L (2018) Pictures from the other side of the fringe: urban growth and peri-urban agriculture in a post-industrial city (Toulouse, France). J Rural Stud 57:25–35

e Silva JDA, Correia M (2023) The main drivers of urban sprawl in Portuguese medium cities between 2001 and 2011. Land Use Policy 132:106803

Egidi G, Cividino S, Vinci S, Sateriano A, Salvia R (2020a) Towards local forms of sprawl: a brief reflection on Mediterranean urbanization. Sustainability 12(2):582

Egidi G, Salvati L, Vinci S (2020b) The long way to tipperary: city size and worldwide urban population trends, 1950–2030. Sustain Cities Soc 60:102148

El Menchawy A, Aly SS, Hakim MA (2011) The impact of urban sprawl on the heritage areas through the urban fabric of cities. WIT Trans Ecol Environ 150:299–314

Escudero Gómez LA, García González JA, Martínez Navarro JM (2019) Medium-sized cities in Spain and their urban areas within national network. Urban Sci 3(1):5

European Environment Agency (2006) Urban sprawl in Europe—the ignored challenge. EEA Report no. 10, Copenhagen

Evers D, Katurić I, van der Wouden R (2024) Urbanization in Europe. Past developments and pathways to a sustainable future. Palgrave Macmillan, London

Feleki E, Achillas C, Vlachokostas C, Michailidou AV, Ortega L, Moussiopoulos N (2018) Preservation of the Mediterranean identity: an intra-city analysis towards a macro-regional approach for the characterisation of urban sustainability. Sustainability 10(10):3551
Feng J, Wang F, Zhou Y (2009) The spatial restructuring of population in metropolitan Beijing: toward polycentricity in the post-reform ERA. Urban Geogr 30(7):779–802
Feng Y, Wang X, Du W, Liu J, Li Y (2019) Spatiotemporal characteristics and driving forces of urban sprawl in China during 2003–2017. J Clean Prod 241:118061
Ferrara A, Kelly C, Wilson GA, Nolè A, Mancino G, Bajocco S, Salvati L (2016) Shaping the role of 'fast' and 'slow' drivers of change in forest-shrubland socio-ecological systems. J Environ Manage 169:155–166
Ferrara C, Carlucci M, Grigoriadis E, Corona P, Salvati L (2017) A comprehensive insight into the geography of forest cover in Italy: exploring the importance of socioeconomic local contexts. Forest Policy Econ 75:12–22
Ferrara A, Kosmas C, Salvati L, Padula A, Mancino G, Nolè A (2020) Updating the MEDALUS-ESA framework for worldwide land degradation and desertification assessment. Land Degrad Dev 31(12):1593–1607
Fuladlu K, Riza M, Ilkan M (2021) Monitoring urban Sprawl using time-series data: Famagusta region of Northern Cyprus. SAGE Open 11(2):21582440211007464
Galster G, Hanson R, Ratcliffe MR, Wolman H, Coleman S, Freihage J (2001) Wrestling sprawl to the ground: defining and measuring an elusive concept. Hous Policy Debate 12(4):681–717
García-Ayllón S (2013) Retrospective analysis of urban development in the Spanish Mediterranean coast. WIT Trans Ecol Environ 179:291–302
García-Coll A, López-Villanueva C (2018) The impact of economic crisis in areas of sprawl in Spanish cities. Urban Sci 2(4):113
Garcia-López MÀ (2019) All roads lead to Rome... and to sprawl? Evidence from European cities. Reg Sci Urban Econ 79:103467
Garcia-López MA, Muñiz I (2010) Employment decentralisation: polycentricity or scatteration? The Case of Barcelona. Urban Stud 47(14):3035–3056
García-Nieto AP, Geijzendorffer IR, Baró F, Roche PK, Bondeau A, Cramer W (2018) Impacts of urbanization around Mediterranean cities: changes in ecosystem service supply. Ecol Ind 91:589–606
Gavalas VS, Rontos K, Salvati L (2014) Who becomes an unwed mother in Greece? Sociodemographic and geographical aspects of an emerging phenomenon. Popul Space Place 20(3):250–263
Giannakourou G (2005) Transforming spatial planning policy in Mediterranean countries: Europeanization and domestic change. Eur Plan Stud 13:319–331
Gielen E, Riutort-Mayol G, Palencia-Jiménez JS, Cantarino I (2018) An urban sprawl index based on multivariate and Bayesian factor analysis with application at the municipality level in Valencia. Environ Plann B Urban Anal City Sci 45(5):888–914
Gonzales S (2011) The north/south divide in Italy and England: discursive construction of regional inequality. Eur Urban Reg Stud 18(1):62–76
Gospodini A (2009) Post-industrial trajectories of Mediterranean European cities: the case of post-Olympics Athens. Urban Stud 46(5–6):1157–1186
Gross J, Ouyang Y (2021) Types of urbanization and economic growth. Int J Urban Sci 25(1):71–85
Grout C, Cavailhès J, Détang-Dessendre C, Thomas A (2016) Is Sprawling residential behavior influenced by climate? Land Econ 92(2):203–219
Habibi S, Asadi N (2011) Causes, results and methods of controlling urban sprawl. Proc Eng 21:133–141
Halbac-Cotoara-Zamfir R, Egidi G, Salvia R, Salvati L, Sateriano A, Gimenez-Morera A (2021) Recession, local fertility, and urban sustainability: results of a quasi-experiment in Greece, 1991–2018. Sustainability 13(3):1052

Halbac-Cotoara-Zamfir R, Marucci A, Salvia R, Quaranta G, Sateriano A, Cecchini M, Bianchini L (2022) Caring of the fringe? Mediterranean desertification between peri-urban ecology and socioeconomics. Sustainability 14(3):1426

Hennig EI, Schwick C, Soukup T, Orlitová E, Kienast F, Jaeger JA (2015) Multi-scale analysis of urban sprawl in Europe: towards a European de-sprawling strategy. Land Use Policy 49:483–498

Hepcan S, Hepcan CC, Kilicaslan C, Ozkan MB, Kocan N (2013) Analyzing landscape change and urban sprawl in a Mediterranean coastal landscape: a case study from Izmir, Turkey. J Coastal Res 29(2):301–310

Hof A, Blázquez-Salom M (2013) The linkages between real estate tourism and urban sprawl in Majorca (Balearic Islands, Spain). Land 2(2):252–277

Ismael HM (2021) Urban form study: the sprawling city—review of methods of studying urban sprawl. GeoJournal 86(4):1785–1796

Ji Y, Guo X, Zhong S, Wu L (2020) Land financialization, uncoordinated development of population urbanization and land urbanization, and economic growth: evidence from China. Land 9(12):481

Kaika M (2003) Constructing scarcity and sensationalising water politics: 170 days that shook Athens. Antipode 35(5):919–954

Kaniewski D, Van Campo E, Morhange C, Guiot J, Zviely D, Shaked I, Artzy M (2013) Early urban impact on Mediterranean coastal environments. Sci Rep 3(1):3540

Karadimitriou N, Maloutas T, Arapoglou VP (2021) Multiple deprivation and urban development in Athens, Greece: spatial trends and the role of access to housing. Land 10(3):290

Kasanko M, Barredo JI, Lavalle C, McCormick N, Demicheli L, Sagris V, Brezger A (2006) Are European cities becoming dispersed? A comparative analysis of fifteen European urban areas. Landsc Urban Plan 77:111–130

Keddie J, Tonkiss F (2010) The market and the plan: housing, urban renewal and socio-economic change in London. City Cult Soc 1(2):57–67

Kemeny T, Storper M (2012) The sources of urban development: wage, housing, and amenity gaps across American cities. J Reg Sci 52(1):85–108

Koprowska K, Łaszkiewicz E, Kronenberg J (2020) Is urban sprawl linked to green space availability? Ecol Ind 108:105723

Kosmas C, Detsis V, Karamesouti M, Kounalaki K, Vassiliou P, Salvati L (2015) Exploring long-term impact of grazing management on land degradation in the socio-ecological system of Asteroussia Mountains, Greece. Land 4(3):541–559

Kosmas C, Karamesouti M, Kounalaki K, Detsis V, Vassiliou P, Salvati L (2016) Land degradation and long-term changes in agro-pastoral systems: an empirical analysis of ecological resilience in Asteroussia-Crete (Greece). CATENA 147:196–204

Kourliouros E (1997) Planning industrial location in Greater Athens: the interaction between deindustrialization and anti-industrialism during the 1980s. Eur Plan Stud 5:435–460

Kovács Z, Farkas ZJ, Egedy T, Kondor AC, Szabó B, Lennert J, Kohán B (2019) Urban sprawl and land conversion in post-socialist cities: the case of metropolitan Budapest. Cities 92:71–81

Kroll F, Kabisch N (2012) The relation of diverging urban growth processes and demographic change along an urban-rural gradient. Popul Space Place 18:260–276

Lagarias A, Prastacos P (2020) Comparing the urban form of South European cities using fractal dimensions. Environ Plann B Urban Anal City Sci 47(7):1149–1166

Lagarias A, Sayas J (2018) Urban sprawl in the Mediterranean: Evidence from coastal medium-sized cities. Reg Sci Inquiry 10(3):15–32

Lagarias A, Sayas J (2019) Is there a common typology of urban sprawl in Mediterranean cities? A comparative analysis utilizing spatial metrics. Revue D'économie Régionale & Urbaine 4:813–850

Lagarias A, Stratigea A (2023) Coastalization patterns in the Mediterranean: a spatiotemporal analysis of coastal urban sprawl in tourism destination areas. GeoJournal 88(3):2529–2552

Le Goix R (2005) Gated communities: sprawl and social segregation in Southern California. Hous Stud 20(2):323–343

Lee S (2011) Metropolitan growth patterns and socio-economic disparity in six US Metropolitan areas 1970–2000. Int J Urban Reg Res 35(5):988–1011
Leontidou L (2009) Beyond the borders of Mediterranean Cities: the Mediterranean City in transition. Quart Int Sociol 18(3–4):131–140
Lindstrom MJ, Bartling H (2003) Suburban sprawl: culture, theory, and politics. Rowman & Littlefield, New York
Liu L, Meng L (2020) Patterns of urban sprawl from a global perspective. J Urban Plann Dev 146(2):04020004
Longhi C, Musolesi A (2007) European cities in the process of economic integration: towards structural convergence. Ann Reg Sci 41:333–351
Martí P, Nolasco-Cirugeda A, Serrano-Estrada L (2017) Assessment tools for urban sustainability policies in Spanish Mediterranean tourist areas. Land Use Policy 67:625–639
Marull J, Pino J, Tello E, Cordobilla MJ (2009) Social metabolism, landscape change and land-use planning in the Barcelona Metropolitan Region. Land Use Policy 27:497–510
Masini E, Tomao A, Barbati A, Corona P, Serra P, Salvati L (2019) Urban growth, land-use efficiency and local socioeconomic context: a comparative analysis of 417 metropolitan regions in Europe. Environ Manage 63:322–337
Mehriar M, Masoumi H, Mohino I (2020) Urban sprawl, socioeconomic features, and travel patterns in Middle-east countries: a case study in Iran. Sustainability 12(22):9620
Millward H (2008) Evolution of population densities: five Canadian cities, 1971–2001. Urban Geogr 29(7):616–638
Morelli VG, Rontos K, Salvati L (2014) Between suburbanisation and re-urbanisation: revisiting the urban life cycle in a Mediterranean compact city. Urban Res Pract 7(1):74–88
Morote ÁF, Hernández M (2016) Urban sprawl and its effects on water demand: a case study of Alicante, Spain. Land Use Policy 50:352–362
Mouratidis K (2019) Compact city, urban sprawl, and subjective well-being. Cities 92:261–272
Muñiz I, Galindo A, Angel Garcia M (2003) Cubic spline population density functions and satellite city delimitation: the case of Barcelona. Urban Stud 40(7):1303–1321
Muñoz F (2003) Lock living: urban sprawl in Mediterranean cities. Cities 20(6):381–385
Musterd S (2005) Social and ethnic segregation in Europe: levels, causes and effects. J Urban Aff 27(3):331–348
Musterd S, Marcińczak S, Van Ham M, Tammaru T (2017) Socioeconomic segregation in European capital cities. Increasing separation between poor and rich. Urban Geogr 38(7):1062–1083
Nickayin SS, Tomao A, Quaranta G, Salvati L, Gimenez Morera A (2020) Going toward resilience? Town planning, peri-urban landscapes, and the expansion of Athens, Greece. Sustainability 12(24):10471
Orenstein DE, Frenkel A, Jahshan F (2014) Methodology matters: measuring urban spatial development using alternative methods. Environ Plann B Plann des 41:3–23
Oueslati W, Alvanides S, Garrod G (2015) Determinants of urban sprawl in European cities. Urban Stud 52(9):1594–1614
Pacione M (2003) Urban environmental quality and human wellbeing—a social geographical perspective. Landsc Urban Plan 65:19–30
Pagliarin S (2018) Linking processes and patterns: spatial planning, governance and urban sprawl in the Barcelona and Milan metropolitan regions. Urban Stud 55(16):3650–3668
Parcerisas L, Marull J, Pino J, Tello E, Coll F, Basnou C (2012) Land use changes, landscape ecology and their socioeconomic driving forces in the Spanish Mediterranean coast (El Maresme County, 1850–2005). Environ Sci Policy 23:120–132
Pardo-García S, Mérida-Rodríguez M (2018) Physical location factors of metropolitan and rural sprawl: geostatistical analysis of three Mediterranean areas in Southern Spain. Cities 79:178–186
Parés M, March H, Saurí D (2013) Atlantic gardens in Mediterranean climates: understanding the production of suburban natures in Barcelona. Int J Urban Reg Res 37:328–347
Paül V, Tonts M (2005) Containing urban sprawl: trends in land use and spatial planning in the metropolitan region of Barcelona. J Environ Planning Manage 48(1):7–35

Perrin C, Nougarèdes B, Sini L, Branduini P, Salvati L (2018) Governance changes in peri-urban farmland protection following decentralisation: a comparison between Montpellier (France) and Rome (Italy). Land Use Policy 70:535–546

Polyzos S, Minetos D, Niavis S (2013) Driving factors and empirical analysis of urban sprawl in Greece. Theor Empir Res Urban Manage 8(1):5–29

Power A (2001) Social exclusion and urban sprawl: is the rescue of cities possible? Reg Stud 35(8):731–742

Pozoukidou G, Ntriankos I (2017) Measuring and assessing urban sprawl: a proposed indicator system for the city of Thessaloniki, Greece. Rem Sens Appl Soc Environ 8:30–40

Pradhan B (2017) Spatial modeling and assessment of urban form. Analysis of urban growth: from sprawl to compact using geospatial data. Springer, Cham

Recanatesi F, Clemente M, Grigoriadis E, Ranalli F, Zitti M, Salvati L (2016) A fifty-year sustainability assessment of Italian agro-forest districts. Sustainability 8(1):32

Rejeb Bouzgarrou A, Claramunt C, Rejeb H (2019) Visualizing urban sprawl effects of a Tunisian city: a new urban spatial configuration of Monastir. Ann GIS 25(1):71–82

Richardson H, Chang-Hee C (2004) Urban sprawl in Western Europe and the United States. Ashgate, London

Robert S, Fox D, Boulay G, Grandclement A, Garrido M, Pasqualini V, Trémélo ML (2019) A framework to analyse urban sprawl in the French Mediterranean coastal zone. Reg Environ Change 19:559–572

Rodrigo-Comino J, Egidi G, Sateriano A, Poponi S, Mosconi EM, Gimenez Morera A (2021) Suburban fertility and metropolitan cycles: insights from European cities. Sustainability 13(4):2181

Rontos K, Grigoriadis S, Sateriano A, Syrmali M, Vavouras I, Salvati L (2016) Lost in protest, found in segregation: divided cities in the light of the 2015 ‘Oki’ Referendum in Greece. City Cult Soc 7(3):139–148

Rontos K, Papazoglou M, Syrmali ME, Sateriano A (2022) Towards the eastern gate to Europe: factors shaping perceptions and attitudes towards migrants in Lesvos Island (Greece), 2016. Reg Stat 12(1):104–134

Rubiera Morollón F, González Marroquin VM, Pérez Rivero JL (2016) Urban sprawl in Spain: differences among cities and causes. Eur Plan Stud 24(1):207–226

Rubiera-Morollón F, Garrido-Yserte R (2020) Recent literature about urban sprawl: a renewed relevance of the phenomenon from the perspective of environmental sustainability. Sustainability 12(16):6551

Salem M, Tsurusaki N, Divigalpitiya P (2020) Land use/land cover change detection and urban sprawl in the peri-urban area of greater Cairo since the Egyptian revolution of 2011. J Land Use Sci 15(5):592–606

Sallustio L, Pettenella D, Merlini P, Romano R, Salvati L, Marchetti M, Corona P (2018) Assessing the economic marginality of agricultural lands in Italy to support land use planning. Land Use Policy 76:526–534

Salvati L (2014) Towards a polycentric region? The socioeconomic trajectory of Rome, an ‘eternal’ Mediterranean city. Tijdschr Econ Soc Geogr 105(3):268–284

Salvati L (2019) Urban dispersion and economic crisis: empirical evidence from a Mediterranean region. J Environ Planning Manage 62(7):1205–1226

Salvati L, Carlucci M (2016) Patterns of sprawl: the socioeconomic and territorial profile of dispersed urban areas in Italy. Reg Stud 50(8):1346–1359

Salvati L, Morelli VG, Rontos K, Sabbi A (2013) Latent exurban development: city expansion along the rural-to-urban gradient in growing and declining regions of southern Europe. Urban Geogr 34(3):376–394

Salvati L, Sateriano A, Saradakou E, Grigoriadis E (2016) ‘Land-use mixité’: evaluating urban hierarchy and the urban-to-rural gradient with an evenness-based approach. Ecol Ind 70:35–42

Salvati L, Ferrara A, Chelli F (2018) Long-term growth and metropolitan spatial structures: an analysis of factors influencing urban patch size under different economic cycles. Geografisk Tidsskrift—Dan J Geogr 118(1):56–71

Salvati L, Ciommi MT, Serra P, Chelli FM (2019) Exploring the spatial structure of housing prices under economic expansion and stagnation: the role of socio-demographic factors in metropolitan Rome, Italy. Land Use Policy 81:143–152

Salvia R, Quaranta V, Sateriano A, Quaranta G (2022) Land resource depletion, regional disparities, and the claim for a renewed 'sustainability thinking' under early desertification conditions. Resources 11(3):28

Sassen S (2010) The city: its return as a lens for social theory. City Cult Soc 1(1):3–11

Saurí D (2003) Lights and shadows of urban water demand management. The case of the metropolitan region of Barcelona. Eur Plann Stud 11(3):229–243

Schneider A, Woodcock CE (2008) Compact, dispersed, fragmented, extensive? A comparison of urban growth in twenty-five global cities using remotely sensed data, pattern metrics and census information. Urban Stud 45:659–692

Schwaab J (2022) Sprawl or compactness? How urban form influences urban surface temperatures in Europe. City Environ Interact 16:100091

Scott AJ, Storper M (2015) The nature of cities: the scope and limits of urban theory. Int J Urban Reg Res 39(1):1–15

Seifollahi-Aghmiuni S, Kalantari Z, Egidi G, Gaburova L, Salvati L (2022) Urbanisation-driven land degradation and socioeconomic challenges in peri-urban areas: insights from Southern Europe. Ambio 51(6):1446–1458

Sharaf SA, Serra P, Saurí D (2018) A district and sector land-use and landscape analysis of urban sprawl in Al Ain municipality (United Arab Emirates): just a quick conversion from sand to a built-up environment? Appl Geogr 95:88–100

Shaw BJ, van Vliet J, Verburg PH (2020) The peri-urbanization of Europe: a systematic review of a multifaceted process. Landsc Urban Plan 196:103733

Soulard CT, Valette E, Perrin C, Abrantes PC, Anthopoulou T, Benjaballah O, Paoli JC (2018) Peri-urban agro-ecosystems in the Mediterranean: diversity, dynamics, and drivers. Reg Environ Change 18:651–662

Storper M, Scott AJ (2016) Current debate in urban theory: a critical assessment. Urban Stud 53(6):1114–1136

Terzi F, Bolen F (2009) Urban sprawl measurement of Istanbul. Eur Plan Stud 17:1559–1570

Tomao A, Quaranta G, Salvia R, Vinci S, Salvati L (2021) Revisiting the 'southern mood'? Post-crisis Mediterranean urbanities between economic downturns and land-use change. Land Use Policy 111:105740

Townsend A (2009) Integration of economic and spatial planning across scales. Int J Public Sect Manag 22(7):643–659

Tsapala F, Chorianopoulos I, Gaki E, Gialis S (2023) The Mediterranean City and EU integration: divergent resurgence paths and new forms of north-south urban divide. Cities 138:104352

Tsilimigkas G, Stathakis D, Pafi M (2016) Evaluating the land use patterns of medium-sized Hellenic cities. Urban Res Pract 9(2):181–203

Tsilimigkas G, Kizos T, Gourgiotis A (2018) Unregulated urban sprawl and spatial distribution of fire events: evidence from Greece. Environ Hazards 17(5):436–455

Turok I, Mykhnenko V (2007) The trajectories of European cities, 1960–2005. Cities 24:165–182

Varela-Candamio L, Rubiera Morollón F, Sedrakyan G (2019) Urban sprawl and local fiscal burden: analysing the Spanish case. Empirica 46(1):177–203

Venanzoni G, Carlucci M, Salvati L (2017) Latent sprawl patterns and the spatial distribution of businesses in a southern European city. Cities 62:50–61

Vidal M, Domene E, Saurí D (2011) Changing geographies of water-related consumption: residential swimming pools in suburban Barcelona. Area 43(1):67–75

Vinci S, Egidi G, López Gay A, Salvati L (2022) Population growth and urban management in metropolitan regions: the contribution of natural balance and migration to polycentric development in Barcelona. Appl Spat Anal Policy 15(1):71–94

Vinci S, Vardopoulos I, Salvati L (2023) A tale of a shrinking City? Exploring the complex interplay of socio-demographic dynamics in the recent development of Attica. Greece. Cities 132:104089

Ward K (2010) Towards a relational comparative approach to the study of cities. Prog Hum Geogr 34(4):471–487

Wolff C, Nikoletopoulos T, Hinkel J, Vafeidis AT (2020) Future urban development exacerbates coastal exposure in the Mediterranean. Sci Rep 10(1):14420

Yılmaz M, Terzi F (2021) Measuring the patterns of urban spatial growth of coastal cities in developing countries by geospatial metrics. Land Use Policy 107:105487

Zambon I, Salvati L (2019) Metropolitan growth, urban cycles and housing in a Mediterranean country, 1910s–2010s. Cities 95:102412

Zambon I, Colantoni A, Carlucci M, Morrow N, Sateriano A, Salvati L (2017) Land quality, sustainable development and environmental degradation in agricultural districts: a computational approach based on entropy indexes. Environ Impact Assess Rev 64:37–46

Zambon I, Benedetti A, Ferrara C, Salvati L (2018) Soil matters? A multivariate analysis of socioeconomic constraints to urban expansion in Mediterranean Europe. Ecol Econ 146:173–183

Zambon I, Colantoni A, Salvati L (2019) Horizontal vs vertical growth: understanding latent patterns of urban expansion in large metropolitan regions. Sci Total Environ 654:778–785

Zwiers M, Bolt G, van Ham M, Van Kempen R (2016) The global financial crisis and neighbourhood decline. Urban Geogr 37(5):664–684

Chapter 2
Kaleidoscopic Morphologies, Socioeconomic Transitions, and the Role of Background Contexts in Mediterranean Urbanities

Abstract Given its unpredictable nature, metropolitan growth in the old continent is considered an intriguing and intricate economic issue at the same time. Urbanization—both compact and dispersed—and suburbanization sequences have occurred in distinctive ways across cities and regions, highlighting peculiar relationships between land use, social dynamics, economic structures, and functions. Settlement dispersion advanced rapidly in Southern Europe—irrespective of city's size and shape—with building rates rising faster than total population. Despite sharing some transformative forces (population dynamics, sociospatial structures, and planning deregulation), urban expansion and metropolitan development have adapted to increasingly volatile economic, cultural, sociodemographic, and environmental contexts. Understanding how place-specific factors influence patterns and processes of metropolitan expansion in Mediterranean contexts may inform policies promoting settlement containment, land-use management, and sustainable development of contemporary cities.

Keywords Urban form · Land take · Economic structure · Social functions · Europe

2.1 Introduction

Metropolitan regions in the Northern Mediterranean basin have represented (and still represent, in most cases) pivotal nodes in a trading system that exceeded the borders of countries. These systems were organized over broader spaces to reflect increasingly complex spatial interactions among economic agents and uncovered geo-political transitions (Allen et al. 2004). Investigating the recent urbanization-suburbanization cycle in Southern Europe allows a coherent description of the main morphological traits shared among regions, in turn outlining distinctive socioeconomic functions and settlement elements at the base of metropolitan transitions (Buzar et al. 2007). In a process of physical expansion and wealth accumulation common to several Mediterranean cities, the uneven role of geographical location and the unrested flow of goods and people have taken the ground—promoting changes that depend on the

L. Salvati, *Economic Growth, Regional Disparities, and Urban Sprawl*, Springer Geography, https://doi.org/10.1007/978-3-031-85799-7_2

intrinsic characteristics of local contexts (Bruegmann 2019). Many of these cities act as central nodes with a wide decision-making power, forming sometimes a 'real-world' economy within the respective national system. Rapid processes of capital accumulation, the hierarchical scale of territories associated with 'the thickness of medium- and long-range relations', as well as the commercial influence and significance of cultural movements distinguished these cities from other production nodes in the region (Angel et al. 2011).

Based on these assumptions, the present chapter attempts at identifying the specificity of Mediterranean urbanization, highlighting its unique origins, different influencing factors and, consequently, varied outcomes—from both morphological and functional perspectives. A general description of long-term urbanization patterns and processes in Southern Europe, as compared with more general trends observed in the old continent, may provide the most appropriate knowledge ground to a more specific description and interpretation of sprawl dynamics in the Mediterranean area. Following the approach proposed in Chap. 1, this rationale appears in line with a basic assumption arguing that any 'general' definition of sprawl may reflect the difficulty of identifying and measuring this phenomenon in vastly different metropolitan realities. Complexity, multidimensionality, and nonlinearity result from some basic characters of sprawl including urban expansion rates, population density, spatial geometry of settlements, accessibility, and aesthetic traits, as clearly highlighted in Egidi et al. (2020).

If a secularist vision has shed light on a stereotypical interpretation of Mediterranean spaces as eminently rural and economically disadvantaged, this approach also reflected an alleged uniformity of metropolitan landscapes (Antrop 2004). Trying to highlight stability and persistency of traditional development paths, the largest Mediterranean cities marked a potential discontinuity in rural landscapes, resulting in a sort of 'scale leap' (Bueno-Suarez and Coq-Huelva 2020). At the same time, they can be interpreted as nodes of global economic networks, connecting the multiple lines of social interaction, economic distribution, and redistribution (Fig. 2.1). They finally reflect changes in the spatial concentration of population, as well as in social institutions and economic functions (Bijker and Haartsen 2012).

Moving from some 'general' definitions of sprawl delineated and discussed earlier in Chap. 1, an additional object of this chapter is a thorough reflection on low-density exurban development adopting a narrative perspective, possibly aided by the visual support of landscape photographs illustrating particular socioeconomic contexts in large metropolitan regions of the Northern Mediterranean basin (Brueckner 2000). This narrative may clarify, likely better than tedious definitions or rigid indicators from statistical sources, how urban sprawl and Mediterranean cities should be taken as 'hegemonic' notions, exactly as they are enucleated and treated widely in the current literature (Brenner and Schmid 2014). The former concept is one of the most significant development modes of Northern American cities—while it was much less debated, for various disciplinary and academic reasons, in the European literature, at least since the early 2000s (Bertinelli and Black 2004).

The latter notion, thanks to the uniqueness of a certain sociological literature, was a reference 'paradigm' for the intrinsic development of 'Southern' cities until the 1990s

Fig. 2.1 Walking in the European city; the renewed archeological pathway downtown in winter Athens. *Source* Authors' photographic archive

(Chorianopoulos et al. 2010). Expressing a latent criticism to the classic Anglo-Saxon urbanization models, the 'romantic' concept referring to the 'Mediterranean city' archetype, introduced a radical change in the notion of 'centers' and 'peripheries', involving a thorough rethinking of what is eminently local, and what is surely global (Chen and Partridge 2013). Our contribution reflects the fascinating interest in both issues, emphasizing the impact that a post-modern interpretation of Mediterranean cities has exerted on a certain geographical literature, even with all its limitations (e.g. Carlucci et al. 2022).

Another aspect highlights the importance of a socioeconomic focus on specific urbanization patterns and processes, allowing a specific interpretation of sequential urbanization-suburbanization waves in Southern cities—as opposed to generic narratives, sometimes mawkish and overly metaphorical (Cassiers and Kesteloot 2012). These elements, sought in the demographic and social traits of local communities and the spatial distribution of economic activities, allow a direct and explicit characterization of low-density settlement expansion, as compared with other types of growth (Table 2.1). The inherent comparison with North American cities as well as with Northern, Western, and Central European urbanities, will contribute to the identification of socioeconomic forces and the reference context that fuel sprawl in complex environments such as the Mediterranean region (Cuadrado-Ciuraneta et al. 2017).

Table 2.1 A generalized strength-weaknesses-opportunities-threats (SWOT) analysis for Mediterranean cities and the related metropolitan regions

Strengths	Weaknesses
Rapid economic development	'City' as a poorly defined notion
Similar demographic trends	Bureaucratic administrative structures
Recovering a symbolic centrality	Modest participation to policy decisions
The supremacy of the 'family'	A new city-suburbs dualism
Unique centers, mixed fringes	Planning deregulation/informality
Opportunities	Threats
Historical architecture	Land take
Pervasiveness of social relations	Disappearance of local cultures
Cultural-economic endowments	Social segregation
Divergence to top-down planning	Convergence to 'westernized' planning
Social permeability of the fringe	Policy and planning centralization
Policies adapted to local contexts	Degraded hinterlands, without identity

Source Author's own elaboration from the literature review presented in this chapter

A mainstream literature in urban studies faced with the apparent similarities between Mediterranean cities, especially investigating settlement form and suburban landscapes since World War II (e.g. Andreotti et al. 2013). Despite some notable exceptions, building compactness was indicated as the peculiar characteristic of many Southern cities (Fig. 2.2), when compared with Northern European agglomerations (Arapoglou and Sayas 2009). These morphologies, however, were losing relevance in recent years, following a continuous dispersion of residential settlements (Andersen et al. 2011). With this context in mind, we started discussing the outcome of such a regional process and its peculiarities in the Mediterranean region (e.g. Alexandri 2015), commenting on the possible socioeconomic and ecological consequences.

Being derived from an extensive literature review, the empirical findings suggest, on the one hand, to (re)formulate a framework interpreting short-term dynamics of Mediterranean cities that can be generalized to other areas with similar characteristics (sensu Akkerman 1992; Anas et al. 1998; Arbaci 2008). On the other hand, they offer a novel interpretation related to urban governance, spatial planning, and metropolitan regulation of settlement development (Balta and Atik 2022), with a view to reducing regional disparities in wealth, land/housing prices and infrastructures (Arbaci and Malheiros 2010), while maintaining environmental, cultural and landscape traits often degraded via massive land consumption dynamics (Biasi et al. 2015), especially in ecologically fragile contexts such as the Mediterranean ecosystems (Bianchini et al. 2021).

Fig. 2.2 Compactness and over-densification as a distinctive trait of Mediterranean cities; downtown Athens, Koukaki, Nea Smirni, and Syngrou neighborhoods in a shot taken from Philopappos hill. *Source* Author's photographic archive

2.2 A Role for (Spatial) Scales

The spatial relationship between cities and the surrounding regions reflects a sort of 'power balance' in the Mediterranean arc, basically resulting in explicit (or latent) gravitational processes between 'centers' and 'peripheries'. Going beyond physical (and functional) size, cities have written the role of rent accumulators in their fate (Boussauw et al. 2012). Wealth imbalances followed the geography of disparities in population distribution and growth. Internal and, subsequently, international migrations to cities have mirrored the economic change consequent to the first industrial revolution, with the gradual abandonment of agricultural activities, and the continuous outsourcing of secondary businesses (Colantoni et al. 2017). The transition to advanced services has laid the foundation for the accelerated growth of contemporary metropolitan systems, while truly mature systems seem to be still infrequent over the regional scene (Coisnon et al. 2014). Additionally, a number of metropolitan systems have been filtered and continuously shaped by strategic choices that still weigh on the structure and balance of national states (Coulson et al. 2013), influencing their role in the global economy.

Although experiencing, for many years, diversified development paths because of the predominance of (i) a capitalist economy in the northwestern shore of the

Mediterranean Sea, (ii) a planned economy in South Eastern Europe, and (iii) a colonial economy in Northern Africa, various types of cities took the upper hand in the hierarchy, originally acting as the biggest agricultural markets in the region and, more recently, playing a role as commercial nodes, or intermediate centers of governance and control (Couch et al. 2007). Coastal cities with harbors, especially when associated with the functional role of (regional or country) capital cities, such as Naples in Italy, Barcelona in Spain, or Athens in Greece, have experienced restless processes of urbanization and suburbanization in the last century, with a considerable population growth and uneven expansion of businesses (De Muro et al. 2011; Chorianopoulos et al. 2014; Cuadrado-Ciuraneta et al. 2017).

Metropolitan hierarchies demonstrated to be spatially articulated in Mediterranean regions. An example of nonlinear hierarchies was presented in Fig. 2.3, illustrating the rank-size rule for population size in Greece (using 2021 census data). This curve ordered human settlements from the smallest village (upper left) to the largest city, Athens (lower right). The existence of a curvilinear relationship between population size and settlement rank in Greece is well documented from the results of a smoothing spline, a nonparametric regression technique delineating at least two economic behaviors: on the left side of the graph, the curve is almost horizontal, indicating a moderate increase of population size with rank, typical of rural areas; on the right part of the graph, the curve became quite vertical, delineating a sudden increase of population size with rank, typical of urban areas. The relationship revealed to be relatively stable with changes in the adopted spatial support, i.e. considering larger administrative units such as municipalities as the elementary analysis' domain (left graph), or focusing on a regular lattice 1 km^2 wide (right graph).

Simplifying definitions and classifications of Mediterranean cities is clearly difficult in the context of specific cases, and hardly generalizable to different contexts. Grasping the many facets of Mediterranean urbanities and its relationship with the national states, fragmentation of the different regional governance and local contexts

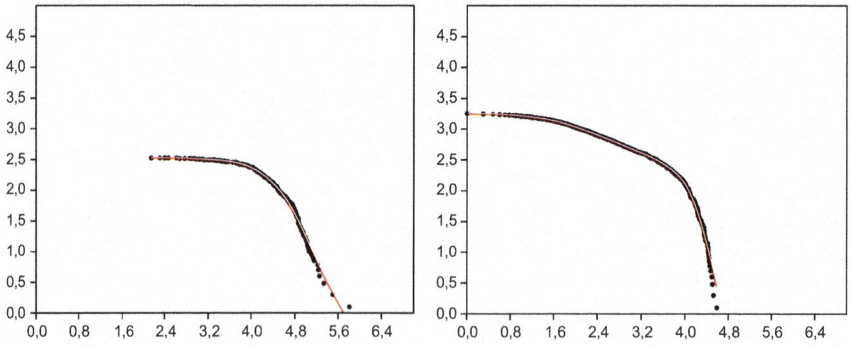

Fig. 2.3 Rank-size rule (log–log scale) of population size (total inhabitants, horizontal axis) against settlement ranking (vertical axis) in Greece (2021 census data) considering 'Kallikratis' NUTS-5 municipalities and a regular population lattice with a 1 km^2 grid. *Source* Author's elaboration on ELSTAT micro-data (left) and Eurostat population grid (right) estimation

requires considering the vast area under study (274 million citizens in the 2000s, divided into 117 million residents along the north bank, and 145 million inhabitants along the south bank). Population increased by three times since the 1970s, when nearly 97 million people resided in the Northern Mediterranean arc and only 50 million people in the Southern bank. The analysis of Mediterranean urban systems clearly highlights the dichotomy between (i) a basically mature urbanization model, while being not completely balanced over space and morphologically dispersed along the Northern bank, and (ii) a more archaic and widely disorganized (despite fully compact) model observed along the Southern shore (De Rosa and Salvati 2016).

Along the Northern shore, metropolitan systems consolidated in recent decades after a relatively long period of intense urbanization. This was a consequence of industrial development, economic growth with the boost of traditional services, and a sudden demographic increase—both because of natural factors (a young population structure with high fertility and low mortality) and migration patterns (the so-called rural–urban exodus first, and growing flows of international immigration, later on). Urban consolidation was progressively realized through a mostly compact expansion of settlements (Fig. 2.4), sometimes alimented by economic processes of spatial decentralization of residences and businesses (Di Feliciantonio and Salvati 2015).

In most cases, this stage has reflected a recovery of population and employment rates in medium-size towns, in line with the sudden development of post-industrial

Fig. 2.4 Spatial sedimentation of additive and compact layers of residential and service settlements in downtown Lisbon in a shot from Gracia. *Source* Author's photographic archive

societies (Durà-Guimera 2003). Urbanization patterns delineate a sustained and continuous growth until the 1970s, and a subsequent stabilization of the share of urban population in total population around 70%, a trend confirmed in the medium-term population projections for the end of the 2020s (Cazcarro et al. 2012). At the national scale, this figure ranges between 65% in Greece and 80% in Spain, being a consolidated trait of Mediterranean societies, although with some local differences (Egidi et al. 2020).

2.3 'Coastalization' as a Form of 'Supra-Regional Landscape'

Urbanization along coastal belts progressed from existing (historical) centers located at (more or less) regular distances and covering the largest part of the Mediterranean arc. The spread of second homes, tourist resorts, shopping malls, service infrastructure, sometimes interspersed with sporadic industrial plants, has contributed to create a mosaic of settlement welding between medium-size centers, stretching inland for several kilometers (Domene and Sauri 2006). This phenomenon evolved in two ways, namely with a gradual incorporation of existing nuclei, or with a tumultuous expansion of spatially differentiated neighboring villages (Domene et al. 2005).

Population concentration along the sea coast was a stylized fact of Mediterranean geographies. Starting in the 1950s along the Northern shore and a bit later along the Southern shore, the process displays some regional facets, because of differential morphologies, topography, and environmental constraints to settlement expansion, similar to what has been observed in geographically distinct contexts (Feng et al. 2019). Night images of the Mediterranean basin with completely built-up coastlines along the North bank, and under gradual expansion along the South bank, are likely regarded as the most eloquent sign of a persistent form of suburban landscapes along the sea coast and increased socioenvironmental complexities inland (Ferrara et al. 2016). According to European Environment Agency (2006) data, residential sprawl was responsible for more than 45% of land transformation into artificial surfaces in flat coastal districts, in turn responding to the increased demand for investments in coastal residences because of tourism development and a rising leisure sector, with visitors mainly from Northern Europe (Fig. 2.5).

Thirty years after the end of the massive 'urbanization' stage in Southern Europe, a progressive slowdown of settlement expansion along the sea coasts was observed, reflecting consolidated linear settlements along the Eastern shore, and the increased importance of productive settlements in the Western shore—penetrating inland and accelerating the infrastructure of rural, disadvantaged and mostly peripheral districts (Ferrara et al. 2014). A gradual demographic saturation of Southern shores was also observed, although a mature organization of these settlements was far from being reached at that time (Garcia-Lopez 2019).

Fig. 2.5 Coastal development of residential, tourism, and second-home settlements in La Cullera, Southern Valencia Province. *Source* Author's photographic archive

Since the late 1980s, population density in Mediterranean Europe reflected coastal spaces that have reached their carrying capacity (Fuhrmann 2020). Northern shores had Italy as the most densely populated country along the sea coast, almost four hundred inhabitants per square kilometer—a value twice the country average. This figure was higher than in Spain, while maintaining a remarkable difference with the national density (Haase et al. 2010). However, density was even higher in the Southern shores: Egypt and Israel were the countries with the highest coastal density, more than five thousand inhabitants per square kilometer, with even more pronounced differentials with grand population density (Gospodini 2009). Density differentials amplified the uneven disparities in rural and urban regions (Gonzales 2011).

Supra-national agglomerations were recognized from Portugal to Turkey: the most evident conurbations include (i) coastal settlements from the Spanish region of Murcia to the West, moving toward Catalonia, Provence, and Liguria to the East, (ii) the linear conurbation along the Adriatic sea coast from Abruzzo to Friuli in Italy, expanding slightly into Croatia; (iii) continuous settlements along the sea coast of Israel, Lebanon and Turkey, up to Egypt, with the even expanding metropolitan region of Alexandria. The coastal conurbation running from Italy (Genoa) to France (Marseille, Montpellier) and Spain (Barcelona-Murcia) is likely the most classical example of Mediterranean linear cities. Following the definition of 'megalopolis' suggested by Gottmann (Vicino et al. 2007), these settlements expanded and elongated between Marseille and Leghorn in a sort of Mediterranean 'teniapolis' (Fielding 1982). This kind of coastal megalopolis includes highly specialized urban functions (Galster et al. 2001), convoluted morphologies (Fotheringham et al. 1989) and mixes residence with industry, as well as historical centers with natural landscapes (Fujita and Mori 1997)—being especially rich in services characterizing it as a proximal example of 'post-industrial' Mediterranean development (Salvati 2022).

Urban fragmentation in coastal districts reveals, since the 1960s, the progressive decline of traditional rural economies along the Northern bank (Garcia-Lopez and Muniz 2010) and, most recently, along the Southern shores (Gkartios 2013). It reflects a sudden crisis of traditional lifestyles, an expanding labor productivity, growing income, and land property, a progressive openness to international trade and gradual integration into the world economy. As suggested in the interpretative framework illustrated in Fig. 2.6, the amplification of coastalization effects on the environment was in some ways associated with the expansion of tourism—a powerful factor of both urbanization and suburbanization typical of the life cycle of any city or region (Gonzales-Val and Olmo 2015). Landscape homologation in coastal districts is thus a direct consequence of such dynamics (Grout et al. 2016).

When locating land capital at the very center of a system of territorial monitoring and conservation policies, it became immediately clear how physical environments

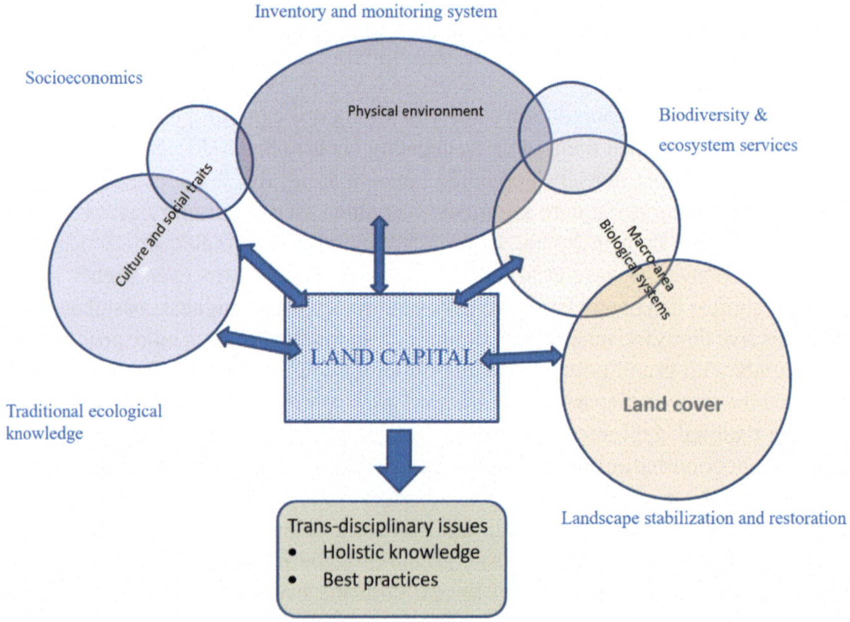

Fig. 2.6 Land capital, sustainable management of fringe districts, and factors of urban fragmentation in metropolitan regions, with specific reference to Southern Europe. *Source* Author's elaboration

are a part of a broader system, where land cover change, cultural and social traits of resident populations, and biodiversity/ecosystem services, are the intrinsic dimensions that make the geography of metropolitan regions in the Mediterranean basin particularly complex. For instance, along the Andalusian sea coast—one of the areas with the highest density of tourist destinations in the Mediterranean, including the renowned centers of Marbella and Torremolinos—land take for buildings has been rapid and continuous, shifting from 37% (in total landscape) in 1975 to 68% in 1990, to the detriment of other land-use, basically cropland and forests (Halbac-Cotoara-Zamfir et al. 2022). To modernize the accommodation capacity and offer new services to the local market, a progressive redevelopment of tourism sites through massive investments in infrastructure has, sometimes, generated new landscapes (Ismael 2021). Changes that occurred along the Adriatic coastal landscape at the turn of the 1980s and the 1990s (due to the increasing specialization of the Opatija Riviera and the tourism destinations in Veneto and Romagna) are common examples of shaping metropolitan hierarchies, re-articulating growth poles, and redesigning residential settlements inland (sensu Hall 1997).

Another factor underlying urbanization processes in coastal areas is the renewed interest in port economies and the rehabilitation of waterfronts (Jacobs-Crisioni et al. 2014). The examples for Barcelona, Valencia, Genoa, Athens, and Thessaloniki, are significant in this regard if intended as a sort of architectonical laboratories trying to find the intrinsic 'mood' of each individual city as lying in the relationship with the sea coast (Kaika 2003). Under such experimentations (Fig. 2.7), these cities were forced to reinvent, at least in part, their heritage, in a context of complete reorganization of maritime trade (Kallis 2010).

These dynamics have also been heavily hit island systems. Tourism development, urban regeneration, and waterfronts, but also land speculation and deregulated spatial planning, will continue playing a role in the economic growth of these areas (Koprowska et al. 2020). Specific instances of a recurring trend that expands from northern to southern regions were traditionally found in the recent development of Balearic and Cycladic Islands (Fig. 2.8). The risk of an uncontrolled urbanization of sea coasts at the expenses of inland territories claims for a rethinking of trans-boundary planning instruments and greater cooperation among local communities and regional authorities (Lampinen et al. 2021). Human capital as harbinger of a proper development, and conservation of natural, agricultural, and cultural heritages, became a basic element of any common action controlling land use and promoting an economic growth tuned finely with social and political development (Chorianopoulos et al. 2010).

Fig. 2.7 'Informal' waterfront of Athens' Riviera in a shot taken from Alimos. *Source* Author's photographic archive

Fig. 2.8 A particularly convoluted landscape sequence from coastal to inland districts in Ornos, Mykonos, one of the most heavily compromised district in the island thanks to intense tourism-driven sprawl. *Source* Author's elaboration courtesy of Google Earth imagery, June 2023

2.4 Toward a New Geography of Mixed (Urban–Rural) Landscapes?

Settlement compactness has played a role in the (re)structuring of contemporary Mediterranean spaces. Demographic polarization in high-density and low-density districts has shaped communities with huge economic disparities and is grounded on hegemonic relationships reflected into metropolitan hierarchies hard to rebalance (Laidley 2016). The divergence between metropolitan areas and the immediate surroundings has brought to the consolidation of (economic and cultural) hierarchical relations among locations (Le Goix 2005); cities have increasingly controlled the surrounding districts. Urban sprawl between the 1980s and the 2010s exalted such dynamics (Lindstrom and Bartling 2003).

Opposed to sprawl, compact growth concentrated in large and medium-size cities, determining a latent process of settlement compaction and hyper-densification (Keddie and Tonkiss 2010), fueling intense differentials in population density between cities and the countryside. In this perspective, the relationship between population density and distance from downtown can be investigated comparatively in various types of cities as a function of their shape (Kasanko et al. 2006). Different patterns can be identified discriminating spatially balanced (and mostly scattered) settlements, leading to some territorial disparities—from a mono-centric growth model, with important consequences on land-use distribution (Kemeny and Storper 2012). While the former structure is typical of polycentric regions (Kourtit et al. 2014), the latter model mostly refers to the archetype of 'dense cities', a typically land-saving way of urban expansion (Kroll and Kabisch 2012). Although it may produce negative environmental externalities (e.g. pollution, congestion, lack of green spaces), settlement compactness (Fig. 2.9) represents a model of urbanization with a particularly low impact as far as land consumption is concerned.

With urbanization becoming progressively decoupled from the intrinsic development dynamics of the largest cities (Legras and Cavailhès 2016), metropolitan growth embodies an intermediate condition, with density gradients in slow attenuation because of sprawl (Lee 2011). Many cities of Western and Central Europe, and those of Southern and Eastern Europe evolving most rapidly, may fall within that class (Kazemzadeh-Zow et al. 2017). In addition to the morphological point of view, the sharp contrast between towns and rural areas in the Mediterranean basin can be read also in the growing cultural and social divide (e.g. Karadimitriou et al. 2021). A marked segregation of city and countryside roles along the economic hierarchy of a given country made this gap more explicit (Halbac-Cotoara-Zamfir et al. 2021). These roles appear as basically pre-formed and difficult to modulate in Mediterranean countries (Lamonica et al. 2020).

The power relationship between central cities and the surrounding locations resulted to be stably regulated under density hierarchies (Longhi and Musolesi 2007). Urban–rural gradients reflect a sort of land-use segregation and the inherent consolidation of distinctive economic roles for compact and dispersed areas (Magliocca et al. 2012). Urban locations are place of consumption and production of goods

Fig. 2.9 Settlement compactness and spatial congestion of kaleidoscopic morphologies and diversified economic functions in downtown Naples. *Source* Author's photographic archive

and (traditional or advanced) services (Maloutas 2007). Rural locations are basically places of agricultural production and conservation (or reproduction) of natural and aesthetic amenities (Masini et al. 2019). A huge segregation between the two locations is intrinsic in any socioeconomic system of the Mediterranean region (Munoz 2003).

Going toward the stylized facts at the base of the relationship between state and regions, Southern societies have often proposed a more subtle interpretation of the metaphysic relationship between cities and the surrounding countryside (Maloutas 2004), as a sort of urban horizon (Fig. 2.10) and, together, a kaleidoscopic form of the intrinsic limit to the complete understanding of complex Mediterranean geographies. In a dominant 'southern' thinking, cities have oneirically represented as the place of culture and socialization, with the countryside being regarded as territories of 'peasants and ignorance' (Mudu 2006). This sort of prejudice against the countryside, in line with the fine-grained microphysics of cities and rural districts, is clearly inverted in the Anglo-American view, assigning a specific role to rural development processes and making the socioeconomic convergence of peripheral places to central locations possible in most cases (Neuman and Hull 2009).

In a seminal work about Mediterranean geographies, Morelli et al. (2014) introduced, from Kayser's seminal thinking, a sort of 'will to power' characteristic of leading cities, giving rise to intense competition along the metropolitan hierarchy in

Fig. 2.10 Urban visions and the Mediterranean horizon; the dusk of a summer sandy-storm day in the Athens' National Garden, with the Acropolis hill in the background. *Source* Author's photographic archive

single countries and across the whole continent. In this perspective, regionalization processes were re-orienting public policies toward few growth poles, to the detriment of a more diffused network of smaller settlements (Mouratidis 2019). Satellite cities took ground along metropolitan hierarchies, with the main city gradually annexing secondary nodes (Metzger 2000). Vacant spaces surrounding central cities (Fig. 2.11) were filled out continuously in a way that all the interstices that separate them from neighboring towns were progressively saturated with population, settlements, and economic activities (Lopez and Hynes 2003).

Urban belts, extended forms of conurbation, submetropolitan poles, are a possible expression of the hegemonic power of mono-centric cities, being relatively far from network models and a balanced polycentric development (Liu and Meng 2020). Going beyond the rhetoric of regionalism and the peculiarity of local architectures, quality of life, consumption styles, individual and collective behaviors, urban cultures and rural traditions make the possible interpretation of these dynamics rather confusing, while challenging (Makse et al. 1998). Resorting to the 'reductio ad unum' typical of an individualistic, 'Nordic' thinking that contrasts the 'Mediterranean koinonia', is an alleged way to face with the popular dimension of urbanization in Southern Europe (Musterd et al. 2017). However, this interpretation can undergo

Fig. 2.11 Vacant land in fringe Lisbon. *Source* Author's photographic archive

a drastic change when assuming sprawl as the dominant aspect of contemporary urbanization (Musterd 2005).

The latent transition of rural locations from 'agents of food production' (agriculture, livestock, forestry, natural amenities) to a 'passive object' supporting city demand for buildable land is even more evident in Mediterranean cities (Nickayin et al. 2020). Globalization at all spatial scales was intrinsically associated with the decline of the original functions in rural areas (Orenstein et al. 2014). Assuming primary goods can be easily and cheaply acquired from very far land, likely better equipped for large-scale farm production (Pacione 2003), 'globalized' residents in unevenly dispersed cities may consider the preservation of the agricultural base surrounding their living (urban) place as a less topical issue. The same citizens may evaluate preserving rural territories from edification as superfluous and economically disadvantageous (Parr 2014). While producing natural amenities and leisure tourism (e.g. Parés et al. 2013), the opportunities for long-term mobility stimulate tourism fluxes concentrated well outside the boundaries of metropolitan regions (Rontos et al. 2016). At the same time, big pipelines provide cities with the (supposedly unlimited) quantity of water collected far from its surroundings—thus perpetrating a supremacy on the neighboring regions, reflected in the unidirectional flow of natural resources extracted from rural districts (Kallis 2010).

As a consequence, rural spaces in Mediterranean regions were becoming a mirror of urban landscapes (Kaika 2003). Not only affluent residents, but also the 'working poor' may escape the congestion of city centers and the isolation of compact peripheries searching for a sort of 'deconcentration' amenities in the countryside (e.g. Power 2001). Cheap second homes—thanks to the expansion and diffusion of smart-working in both public and private jobs—became to be inhabited not only during the short summer holidays (basically corresponding with August in the 'Mediterranean job calendar'), but instead for longer time periods, during both summer and winter (Razin and Rosentraub 2000). A new landscape has been formed, neither city nor countryside. By escaping congestion, however, these new 'residents of intermediate locations' are creating another congestion type, coinciding with the progressive loss of the (hierarchical) order of rural landscapes and land-use classes close to traditional (and compact) Mediterranean cities (Pili et al. 2017).

2.5 The Evolution of a New 'Urban South'

A flourishing sprawl in Southern Europe coincided with a relatively recent reorganization of cities, being sometimes regarded as a wave of 'counterurbanization with population deconcentration' in the life cycle of the region (e.g. Roberts 1991). Scale 'leap' and metropolization are thus symptoms of territorial dynamics where urban development ceased to be spatially contiguous and began to spillover over larger regions, resulting in extended forms of settlement (both residential and productive), economic deconcentration, and social segregation (Partridge et al. 2008). In this view, Mediterranean cities spreaded over a rich mosaic of morphologies sedimented in the course of a millenary history (Rodrigo-Comino et al. 2021).

This evokes the existence of a Mediterranean 'style' of sprawl reflected in the coexistence of multiple settlement arrays, making the spatial analysis of landscapes even more challenging (Richardson and Chang-Hee 2004). This rationale engages the well-known 'conflict' between planning and informality typical of Mediterranean governance (Salvati and Gargiulo Morelli 2014). Hence urban peripheries, being rather indistinguishable from the countryside for many social traits, started spreading around empty spaces (Rubiera-Morollòn and Garrido-Yserte 2020). Suburbs started 'leaching' into rural surroundings, remaining well connected through road and railway infrastructures, and becoming less distinguishable from the morphological point of view (Salvati et al. 2013). A narrative of Barcelona, Spain, would be enough clear to represent the apology of a so vast and pre-defined process. As a result of a long stage of 'architectural experiments' coinciding with a deep transformation of the city since the early 1990s (Fig. 2.12), Barcelona has gradually oriented the original *clichés* of Southern cities, while losing—at least in part—the charm of social tortuosity, morphological convolution, and socioeconomic unpredictability, typical of Mediterranean spaces (Salvati et al. 2016).

Rome, Italy, is another example of an originally semi-compact urban form that has progressively acquired a more dispersed morphology through a process of settlement 'insularisation'. Metropolitan growth in Rome proceeded along radial axes with respect to the historical center, basically along the road system ideated by ancient

Fig. 2.12 From spatially convoluted to regular landscapes in peri-urban spaces of metropolitan Barcelona in the early 2000s. *Source* Author's elaboration of Google Earth imagery

Romans, consolidating a very unusual morphology, not completely compact, while dense and, in some neighborhoods, hyper-dense (Fig. 2.13). This created a fragmented landscape, involving areas well outside the boundaries of central municipalities, and impairing the distinctive features of rural landscapes, the so-called Agro Romano [Roman countryside], which became increasingly compromised because of economic growth (Salvati 2014). In this sense, what is basically seen as a widespread form of urban space—typical of the post-Fordist city—is traditionally regarded as 'exurbs' in regional science, sociology, and urban design. Having all of these disciplines with the objective of delineating a more effective and detailed governance, these morphologies have introduced new organizations of contemporary cities (Salvati and Sateriano 2015).

With this perspective in mind, the comparative profiles delineated in a still young academic literature investigating the recent development path of Mediterranean metropolises, have drawn attention to distinctive settlement typologies, as a generalization of a multidimensional analysis addressing the kaleidoscopic form-functions relationship in Southern Europe, in line with what has been preliminary discussed in Chap. 1 (Couch et al. 2007). A dominant settlement type includes basically polycentric regions, where the traditional settlement history has evolved through a differential growth of large cities, medium-small towns, and the smaller centers in-between (Salvati et al. 2017). While lacking a clear population hierarchy, they act as relatively large and mostly mature regions, still displaying a density gradient from urban districts to rural countryside. In Italy, for instance, such a typology was observed in the Po Valley—from Turin to the West to Venice to the East and Bologna to the South (the so-called polycentric triangle) and, on a smaller scale, in Apulia region, South Eastern Italy. Lowland was the characteristic topography in both areas (Sallustio et al. 2018).

Metropolitan regions with a traditionally mono-centric structure represented the latter settlement type. Being intrinsically characterized by a strict density gradient from larger to smaller centers, which was reflected in compact and spatially continuous settlements (Salvati et al. 2018), central locations economically dominated the surrounding places, while experiencing a wave of growth associated with the demographic boom of the 1970s, and a more or less intense decline in recent decades (Salvati and Carlucci 2016). Population deconcentration at these locations was

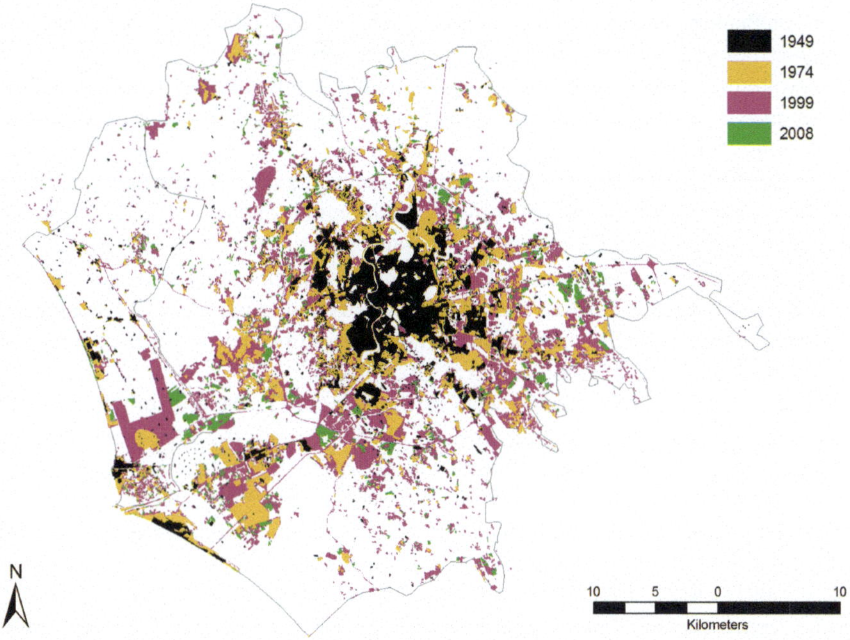

Fig. 2.13 A diachronic map illustrating land-use changes in Rome municipality between 1949 and 2008 (before the great recession). *Source* Author's elaboration of various data sources from national official statistics and international programs of environmental surveillance

frequently intermixed with spatial decentralization of businesses, in turn, reflected in a progressive relocation of industrial polygons in suburban spaces (Scott et al. 2013). Consequently, central business districts with traditional and/or advanced services repositioned frequently along the fringe (Salvati et al. 2019). This settlement type includes large cities such as Rome and Naples in Italy, Athens and Thessaloniki in Greece, as well as Lisbon and Oporto in Portugal. Massive relocation of urban populations to suburbs—especially through processes of social filtering and horizontal segregation—has consolidated this development path (Schuetze et al. 2015).

In this regard, settlement dispersion was interpreted as the consolidation of new forms of living that generate a different spatial organization compared with the traditional models of economic agglomeration and scale, dominant in the modern city (Scott 2008). Moderately dispersed morphologies consolidated in a network of medium-small residential fabrics embedded into low-density housing lattices and extensive areas of traditional (and, more rarely, advanced) services (Scott and Storper 2015). This model reveals a 'weak' urbanization of the new interstitial spaces, well described in a 'territorial' literature interpreting sprawl as a process that breaks the social patterns of a given community (Sauri 2003). New forms of territorial governance are increasingly asked to regulate sprawl and govern the intrinsic metropolitan transformations reflecting the challenge for truly sustainable and resilient systems. These actions will overcome the traditional antinomy between planning and informality, in turn restoring the original nexus between local communities and environmental endowments (Sassen 2010).

Despite the heterogeneity of metropolitan contexts, an attempt to contextualize the stylized facts of Mediterranean sprawl may initially refer to major metropolitan areas of the Northern shore, comparing such features with those observed in other European regions as well as in North America, the area where sprawl was observed and described, likely for the first time in the literature (Salvia et al. 2022). When integrating different reading frames, structural and functional differences can be found at both regional and local scales, distinguishing morphological aspects from socioeconomic processes (Salvati and Serra 2016). Although these differences are predicted to increase in the coming future, it is believed that they have a great influence on exurban development (Storper and Scott 2016).

In accordance with the seminal work by Sinclair (1967), these assumptions apply especially to some composing dimensions of sprawl, both from the perspective of urban form (e.g. compactness, density, shape, convolution, fragmentation, proximity) and social dynamics (e.g. horizontal/vertical segregation, ethnic segregation, economic polarization, business diversification, increased spatial disparities, wealth accumulation). The impact of sprawl on environmental quality was also highlighted, through integrated visions that quantify the role of land consumption and soil sealing in the recent dynamics of metropolitan regions, focusing on e.g. landscape fragmentation, loss of biodiversity, and the progressive disappearance of cultural traits (Smiraglia et al. 2015). Despite a relatively long process of exurban development, per-capita settlement area (m^2), likely one of the most direct indicators of land take and, together, compact urbanization, still depicts Mediterranean countries as the most parsimonious context in the old continent, having consumed less than 500 m^2 of land per inhabitant in 2009 and 750 m^2 of land per inhabitant in 2018 (Fig. 2.14).

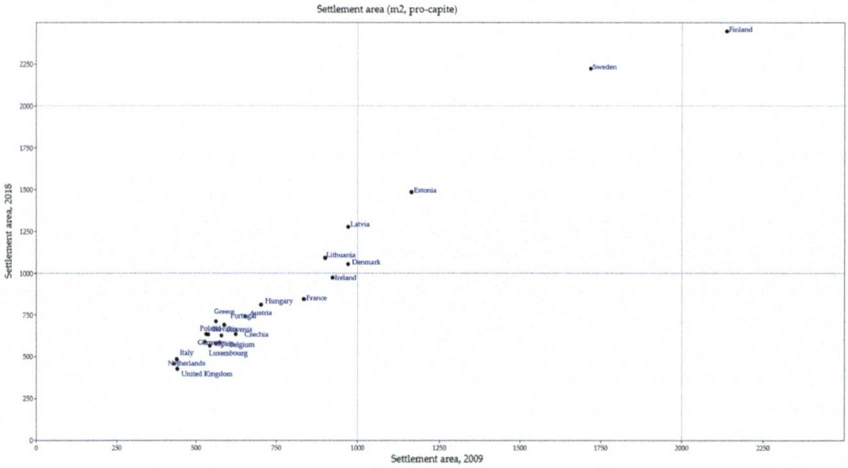

Fig. 2.14 Relationship between settlement area (per-capita m^2) in 2009 (x-axis) and 2018 (y-axis) in selected European countries; *Source* Author's elaboration on Eurostat database

Based on statistical data, other (non-Mediterranean) countries displaying compact or semi-dense settlements in both years include the United Kingdom, the Netherlands, Luxembourg, Germany, and Poland, with France being more oriented toward medium–low density settlements. Keeping around 1000 m^2 per inhabitant at both observation years, Northern European countries distinguished in the sample for a very low settlement density—that maintained stable over time.

2.6 The Mediterranean 'Urban Dream'

When compared with parallel literatures from the United States and other advanced economic systems in Canada, Australia, Japan, and the rest of Europe, sprawl came into the foreground relatively later as a research issue in the geographical studies dealing with Mediterranean urbanities. Three synergistic aspects were especially highlighted in this supposedly young literature:

(i) the impact of different morphologies (originally compact or traditionally dispersed),
(ii) the role of planning deregulation, and
(iii) the intimate contrast between population structure and dynamics, reflecting the intrinsic complexity of metropolitan regions.

For the first dimension, it has been said that the majority of large cities in the Mediterranean basin—both in the Northern shore and in the Southern bank—displayed settlement compactness as a distinctive feature of its landscape (Kasanko et al. 2006). Even some cities grown rapidly in the last century and organized across traditionally branched morphologies (Serra et al. 2014)—with medium-to-high population density and well-developed infrastructural networks, have retained some distinctive characters in the design of residential settlements, being incomparable with the representative Anglo-Saxon and Northern European urban traits (Seifollahi-Aghmiuni et al. 2022).

In structural terms, dense urban expansion reflects a radio-centric growth proceeding additively from downtown to peripheries. In functional terms, urban growth produced less conclusive outcomes, possibly connected with a consolidation of specific socioeconomic dynamics such as population concentration, horizontal/vertical segregation, economic diversification, unequal and spatially unbalanced distribution of public services, as well as heterogeneous access to transport infrastructure (Townsend 2009).

From the ecological point of view, compact growth determined a modest rate of land take (Fig. 2.15), destroying land already degraded and possibly 'waiting' speculatively for building. Vacant land downtown experienced, in such contexts, increasing speculative pressures under volatile real estate markets (Swyngedouw 2004). Social dynamics, however, seem to be an important element crystallizing this kind of metropolitan expansion until the 1980s. A growing demand for residential dwellings consolidated, at least up to the late 1990s, thanks to the emergence of a

Fig. 2.15 Peri-urban agriculture embedded along Rome's fringe: olive groves, vineyards, pastures and shrubland offer an indirect constraint to exurban development in Tivoli countryside. *Source* Author's photographic archive

latent narrative envisaging big cities as a sort of 'promised land'—free from economic restrictions, and the cultural poverty of the countryside (Vidal et al. 2011).

Sociological studies using personal interviews demonstrated, still in the 2000s, how residents of Mediterranean cities have been rarely seeking a sort of 'rural utopia', similar to what has been earlier observed in North America. Sprawl in Southern Europe involved local communities inspired with a geographical imagination of urbanity rather than any sort of 'rural idyll' (Tsilimigkas and Kizos 2014). Thanks to a long history of underdevelopment, 'rurality' in a great part of Mediterranean Europe has been regarded as a synonymous of economic backwardness, poverty, and social insecurity, determining a cogent need to migrate (Vinci et al. 2022).

Urbanities, on the contrary, have been synonymous with economic prosperity (Fig. 2.16), better job opportunities, social amenities, and infrastructure associated with a higher quality of life (Zambon et al. 2019). While large masses of rural populations moved to cities, internal immigrants were sometimes not able to afford huge (and increasing) housing prices downtown and were forced to build-up their settlements as close as possible to historical centers, determining a rising demand for spontaneous settlements (Rontos et al. 2022). The characteristic 'astyphilia' [a Greek term regarded as a sort of cogent demand for urban areas] of this population segment was mainly pragmatic and related to survival strategy. Following Couch et al. (2007), their

Fig. 2.16 'Mediterranean-Atlantic' dream of economic prosperity and social development; a picture of the '25 de Abril' bridge on Tago River taken from Almada and looking toward the north at the vast metropolitan region of Lisbon. *Source* Author's photographic archive

settlement reflected infrastructural development and sought employment directly in urban communities, often entering informal economic sectors, rather than exploiting more formal job opportunities.

An additional aspect refers to the increasing vulnerability of Mediterranean ecosystems to various environmental pressures (including climate aridity and drought, susceptibility to seismic and hydrological risks, massive contamination of soil, water, and air), which is relatively less intense in the major urban areas of Northern and Western Europe (Tomao et al. 2017). In this context, settlement dispersion demonstrated to have a negative impact on land vulnerability, shaping less resilient territories where environmental quality is frequently seconded to economic development (Wilson et al. 2017). In accordance with Zwiers et al. (2016), the comparative analysis of different case studies allows to outline the stylized facts of Mediterranean sprawl:

(i) sprawl is a relatively recent phenomenon in the Mediterranean basin, potentially more aggressive than what has been observed in both Central and Northern Europe;
(ii) this phenomenon held simultaneously across countries at the end of an intense wave of population growth (namely the 'urbanization' stage of the city life

cycle) typical of the 1960s and the 1970s, corresponding with radio-centric and compact development of strictly urban areas;
(iii) the process is undoubtedly multidimensional and pervasive across the three dimensions of sustainability.

From the economic point of view, this phenomenon is part of a background context where building activity (planned or, less frequently, informal) was still a driving force of local economic growth in cities where, however, aspiration to an improved position in the global market is getting high and high (Zambon et al. 2017). As far as the social dimension is concerned, such dynamics impact mainly on processes of horizontal and vertical segregation, typical of Southern cities, reconstituting a complex picture of metropolitan geography through increased demographic polarization and deconcentration of central districts (Tomao et al. 2021). From the environmental point of view, the phenomenon adds some ecological issues typical of European cities, to the large impact of territorial fragility of Mediterranean suburban spaces, the traditional weakness of the agricultural sector, the social permeability of fringe communities, and the economic dependence of rural areas from few urban locations (Sauri 2003). These considerations testify the importance of settlement patterns in the current growth path (Vinci et al. 2023).

Mediterranean urbanity is, finally, set up in a context of weak planning, which has failed to regulate the rapid growth since World War II, taken as a form of local development that often results into a sort of 'sprawl without growth' (Chorianopoulos et al. 2010). Many cities in Southern Europe have shown a progressive convergence toward the North American sprawl 'style'—mainly based on a low-density and spatially discontinuous settlement expansion (Munoz 2003). In this perspective, future research should better investigate how sprawl and compactness effectively represent, for any city, a source of economic strength and socioenvironmental weakness (Carlucci and Salvati 2023). This will explicitly answer the question on what are the economic benefits of sprawl, and the extent to which cities are actually benefiting from such processes (Salvia et al. 2022).

What are the negative feedbacks on the social cohesion of the related regional system is another issue that should be carefully investigated (Arbaci 2008). To answer so articulated questions requires a comparative analysis accounting for the peculiarity of different metropolitan contexts in Southern Europe (sensu Le Goix 2005). Only a generalized comprehension of the (economic-social-ecological) stylized facts associated with sprawl may contribute to design (or redesign) practical solutions of urban containment and sustainable development addressing a latent—but so impactful—process such as exurban development.

2.7 Discussing the Facts, Envisaging the Future

The contemporary growth of Mediterranean cities and the development path characteristic of their surrounding regions have been frequently regarded as a powerful factor in the formation of spatial divides in economic processes and territorial imbalances in social dynamics (Rontos et al. 2016). For instance, social polarization and the abnormal concentration of production activities in some urban areas, combined with a tumultuous population growth, have caused significant disparities with inner regions, often associated with higher environmental risk, land resource depletion, and natural disasters (Fig. 2.17).

Practical solutions, planning instruments, and local development policies demonstrated to adapt only partly to such kinds of metropolitan transition (Serra et al. 2014). The emergence of diseconomies in infrastructure, as well as human and financial capital, has not been dealt with promptly and, in some cases, it also appeared to be institutionally supported in various ways—especially designing traditional and centralized industrial policies (Neuman and Hull 2009). This condition has led to major environmental imbalances, congestion, air, soil and water pollution, environmental fragmentation, and a widespread loss of biodiversity, together with tourism externalities.

Fig. 2.17 Human pressure, settlement coastalization, and forest fires along fringe land in Tindari, Northern Sicily. *Source* Author's photographic archive

Imbalances of the economic functions between central and peripheral locations reflected the structural weakness of the primary sector, with environmental consequences alimenting social conflicts. Land fragility at the fringe was initially addressed through barely technical approaches and local actions with a strong sectoral characterization and limited participation of citizens in planning decisions (Mudu 2006). More recently, the spread of co-operative approaches to sectoral and territorial issues promoted an active involvement of stakeholders working at the local scale.

Convergence to continental European patterns and trends was manifested in Mediterranean regions with the progressive adoption of structural policies aimed at containing land consumption, often in line with some European Commission recommendations, in turn crystallized in several experiences stemming from local initiatives of Agenda 21 (Magliocca et al. 2012). Nevertheless, there are still open issues in both positive and normative aspects of sprawl research:

(i) the loss of agricultural and semi-natural landscapes with high ecological value,
(ii) the disappearance of cultural heritages,
(iii) 'fringe trivialization',
(iv) the failure in containing fringe land degradation, and
(v) the increased biophysical risk following metropolitan development in potentially ineligible areas because of steepness, flooding, or wildfires.

The difficult management of municipal solid waste, water shortage, and the lack of services in media technology (especially in some fringe settlements recently legalized after decades of informality), traffic congestion and the lack of a truly integrated public transport system (Lopez and Hynes 2003), as well as weak policies in combating air pollution, are additional examples of the inherent crisis in sprawled settlements along the Mediterranean region (Kaika 2003).

Ambitious goals can be achieved in the construction of a common cultural and institutional platform for the sustainable management of coastal areas experiencing sprawled urban expansion. Particularly interesting is the promotion of multifunctional agriculture acting as a buffer to exurban development and, together, land degradation in traditional Mediterranean landscapes (Salvati et al. 2017). Such actions can be achieved with a strategic planning of vacant land, considering together political, technical, social, and functional aspects. A continuous exchange of best practices, sharing common goals for sustainable development, introducing co-financing mechanisms that support practical actions on the field, and considering penalty instruments in the case of abuse or misuse of land, are operational tools that can be included in any strategy of sustainable fringe management (Liu and Meng 2020).

Finally, such strategies should incorporate measures of suburban redevelopment, waste management, recovery and upgrade of abandoned industrial and commercial spaces, with the final aim at reverting the economic dynamics of cities and regions at the edge of the global scene. The first monitoring stage should include a thorough assessment of biophysical and monetary depletion of soil resources (Fig. 2.18), regarded as a preliminary step to any study dealing with land degradation and desertification risk, especially along fringe districts.

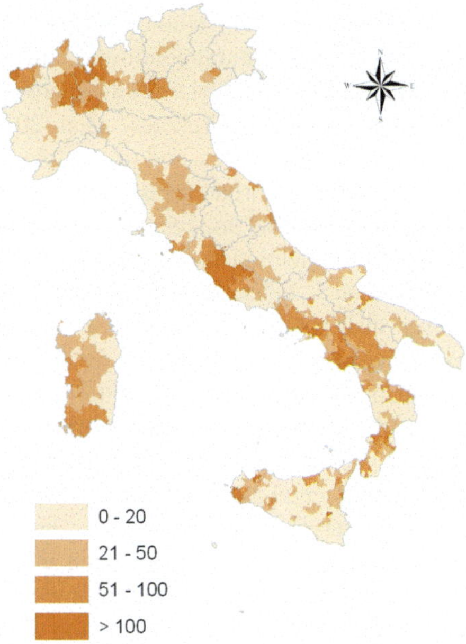

Fig. 2.18 Geography of land depletion in Italy estimating the monetary decline of soil resources (euros per-capita); source: author's elaboration on various national and international geo-spatial databases

Planning local development processes in the Mediterranean region is challenging, and past experiences, at least in some cases, help understanding the intrinsic complexity of socioeconomic phenomena (Salvati et al. 2016). Trade interconnections, as well as promiscuity of peoples, cultures, and traditions fascinating historians for a long time, have contributed to the definition of a (only supposed) homogeneous region where the sea was the preferred medium of trade and cultural exchange, creating a common landscape for planning, local development, sociodemographic policies, and land management at large (Domene and Sauri 2006). Solutions borrowed from allochthonous cultures, and operational actions (planning/policy/management) derived from distant experiences and from a sort of 'functionalist' approach less grounded in local realities that, without the necessary social permeation, failed in providing effective answers to all these complex issues (Domene et al. 2005).

In line with the 'global style' permeating any transformation of urban spaces and metropolitan landscapes, the lack in a local understanding of architectonical practices characterized recent changes in the heart of Mediterranean cities. Motivated in the name of universal principles of 'modern urbanism' (e.g. Hall 1997), they have progressively reduced the image of the Mediterranean cities to a sort of artistic ('museum') places, consolidating the spatial fracture with rural architectures in the surrounding landscapes, considered as 'immature' metropolitan landscapes—and not as pre-existing realities that characterize peri-urban spaces with complex and diversified features (Gospodini 2009).

Mediterranean urban studies, whose results should be seen as the effect of articulated social and cultural conditions, still represent the expression of deeply stratified conditions—which are not the mere consequence of an economic 'delay' with 'Northern' countries (Grekousis et al. 2013). Planning practices and development policies reflecting models and methodologies culturally distant and decoupled from the Mediterranean urban transformations, seem to have raised very modest results toward sustainability and resilience (Gkartios 2013).

While urbanization has been closely related to economic development and demographic change, heterogeneous patterns and processes of regional growth reflect the uneven distribution of urbanization, the subtle impact of demographic dynamics, and the consequent implications for land resource management and environmental sustainability. Differences in patterns of urban growth in a paradigmatic region such as the Mediterranean basin—often masked by official statistics indicating a net increase in urban population—result in territorial divides in some sociodemographic and environmental aspects. An integrated rethinking of the issue should be oriented toward the search for new definitions and interpretations, especially proposing original reading frames of Mediterranean sprawl. Outlining the role and impact of local contexts, is an important aspect of any definition of sprawl as a joint type of space, settlement, or landscape, in any region of the world.

A refined analysis of the impact diversified socioeconomic contexts have on sprawl patterns and processes may also improve reliability, spatial accuracy, and temporal precision of land-use projections. These forecasts reveal to be meaningful when predicting the future growth of cities and metropolitan regions at large. In this regard, geo-referenced databases mixed with local-scale narrative (and visual) information possibly covering medium-long time intervals, are essential to provide the basic knowledge required to identify spatial regimes of urban growth, the influence of population density, the role of economic dynamics, and the impact of planning regulations. The integration of basic socioeconomic indicators made available over sufficiently long time periods, with empirical evidence from a literature review, in situ (e.g. photographic) narratives, and field surveys providing primary data, will contribute to a comparative analysis of the form-function relationship at the base of metropolitan dynamics, and may also clarify the role of local contexts when designing and implementing joint strategies for spatial planning and sustainable regional development.

2.8 Concluding Remarks

While preserving regional systems that are still far from economic maturity, Mediterranean spaces are becoming progressively more heterogeneous because of divergent environmental processes and building dynamics they are embarking in. Despite sprawl, landscape production—in both cities and regions—continues to be associated with the sea coastline, having in the historical towns and the interior (agricultural) plains, the territorial nodes of particularly complex and interacting forms and functions. Insisting on the economic opportunities inherent in coastal territories, the recent production of urban spaces was still based on spatially explicit processes of 'coastalization', determining huge socioenvironmental problems, still not completely addressed.

Reconnecting the impacts of regional-scale socioeconomic change with local-scale ecological dynamics definitely contributes to an enriched knowledge of environmental histories, outlining how a study of differences under assumptions of nonlinear dynamics and 'complex system thinking' is key to understand future socioenvironmental trends in Southern Europe. With this perspective in mind, moving from continental/country to regional/local levels of investigation, a multidimensional, diachronic analysis of socioenvironmental change in metropolitan regions gives key opportunities to ascertain the combined economic and ecological impact of global change. This also outlines the role of social transformations and the latent linkage with 'hegemonic' concepts such as land take and soil resource depletion, which are intimately related with both socioeconomic dynamics and environmental sustainability.

Based on empirical findings, the present chapter encourages research within a multidisciplinary (theoretical and operational) space, stimulating a discussion of integrated issues in-between economics and environmental science, and proposing new empirical frameworks for complex systems' analysis. Practical approaches, comparative works, and case studies are expected to provide the necessary, informed ground to both science and policy. Finally, policies at varying spatial and operational scales are increasingly required to play a regulation role in the construction of coastal landscapes, in line with the intrinsic characteristics of Mediterranean urbanities.

References

Akkerman A (1992) Fuzzy targeting of population niches in urban planning and the fractal dimension of demographic change. Urban Stud 29(7):1093–1113

Alexandri G (2015) Reading between the lines: gentrification tendencies and issues of urban fear in the midst of Athens' crisis. Urban Stud 52(9):1631–1646

Allen J, Barlow J, Maloutas T, Padovan L (2004) Housing in southern Europe. Blackwell, London

Anas A, Arnott R, Small KA (1998) Urban spatial structure. J Econ Lit 36(3):1426–1464

Andersen HT, Møller-Jensen L, Engelstoft S (2011) The end of urbanization? Towards a new urban concept or rethinking urbanization. Eur Plan Stud 19(4):595–611

Andreotti A, Le Galès P, Moreno Fuentes FJ (2013) Controlling the urban fabric: the complex game of distance and proximity in European upper middle class residential strategies. Int J Urban Reg Res 37(2):576–597

Angel S, Parent J, Civco DL, Blei A, Potere D (2011) The dimensions of global urban expansion: estimates and projections for all countries, 2000–2050. Prog Plan 75(2):53–107

Antrop M (2004) Landscape change and the urbanization process in Europe. Landsc Urban Plan 67:9–26

Arapoglou VP, Sayas J (2009) New facets of urban segregation in Southern Europe. Eur Urban Reg Stud 16(4):345–362

Arbaci S (2008) (Re)Viewing ethnic residential segregation in southern European cities: housing and urban regimes as mechanisms of marginalisation. Hous Stud 23(4):589–613

Arbaci S, Malheiros J (2010) De-segregation, peripheralisation and the social exclusion of immigrants: southern European cities in the 1990s. J Ethn Migr Stud 36(2):227–255

Balta S, Atik M (2022) Rural planning guidelines for urban-rural transition zones as a tool for the protection of rural landscape characters and retaining urban sprawl: Antalya case from Mediterranean. Land Use Policy 119:106144

Bertinelli L, Black D (2004) Urbanization and growth. J Urban Econ 56(1):80–96

Bianchini L, Egidi G, Alhuseen A, Sateriano A, Cividino S, Clemente M, Imbrenda V (2021) Toward a dualistic growth? Population increase and land-use change in Rome, Italy. Land 10(7):749

Biasi R, Colantoni A, Ferrara C, Ranalli F, Salvati L (2015) In-between sprawl and fires: long-term forest expansion and settlement dynamics at the wildland–urban interface in Rome, Italy. Int J Sust Dev World 22(6):467–475

Bijker R, Haartsen T (2012) More than Counter-urbanisation: migration to popular and less-popular rural areas in the Netherlands. Popul Space Place 18:643–657

Boussauw K, Neutens T, Witlox F (2012) Relationship between spatial proximity and travel-to-work distance: the effect of the compact city. Reg Stud 46(6):687–706

Brenner N, Schmid C (2014) The 'urban age' in question. Int J Urban Reg Res 38(3):731–755

Brueckner JK (2000) Urban sprawl: diagnosis and remedies. Int Reg Sci Rev 23(2):160–171

Bruegmann R (2019) Sprawl: a compact history. University of Chicago Press, Chicago

Bueno-Suárez C, Coq-Huelva D (2020) Sustaining what is unsustainable: a review of urban sprawl and urban socio-environmental policies in North America and Western Europe. Sustainability 12(11):4445

Buzar S, Ogden P, Hall R, Haase A, Kabisch S, Steinfiihrer A (2007) Splintering urban populations: emergent landscapes of reurbanisation in four European Cities. Urban Stud 44(4):651–677

Carlucci M, Salvati L (2023) Assessing path-dependent urban growth with geographically weighted regressions. Environ Impact Assess Rev 98:106920

Carlucci M, Vinci S, Ricciardo Lamonica G, Salvati L (2022) Socio-spatial disparities and the crisis: swimming pools as a proxy of class segregation in Athens. Soc Indic Res 161(2):937–961

Cassiers T, Kesteloot C (2012) Socio-spatial Inequalities and Social Cohesion in European Cities. Urban Stud 49(9):1909–1924

Cazcarro I, Duarte R, Sánchez-Chóliz J (2012) Economic growth and the evolution of water consumption in Spain: a structural decomposition analysis. Ecol Econ 96:51–61

Chen A, Partridge MD (2013) When are cities engines of growth in China? Spread and backwash effects across the urban hierarchy. Reg Stud 47(8):1313–1331

Chorianopoulos I, Pagonis T, Koukoulas S, Drymoniti S (2010) Planning, competitiveness and sprawl in the Mediterranean city: the case of Athens. Cities 27(4):249–259

Chorianopoulos I, Tsilimigkas G, Koukoulas S, Balatsos T (2014) The shift to competitiveness and a new phase of sprawl in the Mediterranean city: enterprises guiding growth in Messoghia—Athens. Cities 39:133–143

Coisnon T, Oueslati W, Salanié J (2014) Urban sprawl occurrence under spatially varying agricultural amenities. Reg Sci Urban Econ 44:38–49

Colantoni A, Grigoriadis E, Sateriano A, Sarantakou E, Salvati L (2017) Back to Von Thunen: a southern European perspective on mono-centric urban growth, economic structure and non-urban land decline. Int Plan Stud 22(3):173–188

Couch C, Petschel-Held G, Leontidou L (2007) Urban sprawl in Europe: landscapes, land-use change and policy. Blackwell, Oxford, UK

Coulson NE, Liu CH, Villupuram SV (2013) Urban economic base as a catalyst for movements in real estate prices. Reg Sci Urban Econ 43(6):1023–1040

Cuadrado-Ciuraneta S, Durà-Guimerà A, Salvati L (2017) Not only tourism: unravelling suburbanization, second-home expansion and "rural" sprawl in Catalonia, Spain. Urban Geogr 38(1):66–89

De Rosa S, Salvati L (2016) Beyond a 'side street story'? Naples from spontaneous centrality to entropic polycentricism, towards a 'crisis city.' Cities 51:74–83

De Muro P, Monni S, Tridico P (2011) Knowledge-based economy and social exclusion: shadow and light in the Roman socio-economic model. Int J Urban Reg Res 35(6):1212–1238
Di Feliciantonio C, Salvati L (2015) 'Southern' alternatives of urban diffusion: investigating settlement characteristics and socio-economic patterns in three Mediterranean Regions. Tijdschr Econ Soc Geogr 106(4):453–470
Domene E, Sauri D (2006) Urbanization and water consumption: influential factors in the metropolitan region of Barcelona. Urban Stud 43(9):1605–1623
Domene E, Sauri D, Parés M (2005) Urbanization and sustainable resource use: the case of garden watering in the metropolitan region of Barcelona. Urban Geogr 26(6):520–535
Durà-Guimera A (2003) Population deconcentration and social restructuring in Barcelona, a European Mediterranean city. Cities 20(6):387–394
Egidi G, Salvati L, Vinci S (2020) The long way to tipperary: city size and worldwide urban population trends, 1950–2030. Sustain Cities Soc 60:102148
Feng Y, Wang X, Du W, Liu J, Li Y (2019) Spatiotemporal characteristics and driving forces of urban sprawl in China during 2003–2017. J Clean Prod 241:118061
Ferrara A, Salvati L, Sabbi A, Colantoni A (2014) Urbanization, soil quality and rural areas: towards a spatial mismatch? Sci Total Environ 478:116–122
Ferrara A, Kelly C, Wilson GA, Nolè A, Mancino G, Bajocco S, Salvati L (2016) Shaping the role of 'fast' and 'slow' drivers of change in forest-shrubland socio-ecological systems. J Environ Manage 169:155–166
Fielding AJ (1982) Counterurbanization in western Europe. Prog Plan 17:1–52
Fotheringham AS, Batty M, Longley PA (1989) Diffusion-limited aggregation and the fractal nature of urban growth. Pap Reg Sci 67(1):55–69
Fuhrmann M (2020) Port cities of the eastern Mediterranean: urban culture in the late Ottoman Empire. Cambridge University Press, Cambridge, UK
Fujita M, Mori T (1997) Structural stability and evolution of urban systems. Reg Sci Urban Econ 27(4–5):399–442
Galster G, Hanson R, Ratcliffe MR, Wolman H, Coleman S, Freihage J (2001) Wrestling sprawl to the ground: defining and measuring an elusive concept. Hous Policy Debate 12(4):681–717
Garcia-López MÀ (2019) All roads lead to Rome… and to sprawl? Evidence from European cities. Reg Sci Urban Econ 79:103467
Garcia-López MA, Muñiz I (2010) Employment decentralisation: polycentricity or scatteration? The case of Barcelona. Urban Stud 47(14):3035–3056
Gkartios M (2013) 'Leaving Athens': narratives of counterurbanisation in times of crisis. J Rural Stud 32:158–167
Gonzales S (2011) The north/south divide in Italy and England: discursive construction of regional inequality. Eur Urban Reg Stud 18(1):62–76
Gonzales-Val R, Olmo J (2015) Growth in a cross-section of cities: location, increasing returns or random growth? Spat Econ Anal 10(2):230–261
Gospodini A (2009) Post-industrial trajectories of Mediterranean European cities: the case of post-Olympics Athens. Urban Stud 46(5–6):1157–1186
Grekousis G, Manetos P, Photis YN (2013) Modeling urban evolution using neural networks, fuzzy logic and GIS: the case of the Athens metropolitan area. Cities 30:193–203
Grout C, Cavailhès J, Détang-Dessendre C, Thomas A (2016) Is sprawling residential behavior influenced by climate? Land Econ 92(2):203–219
Haase A, Kabisch S, Steinführer A, Bouzarovski S, Hall R, Ogden P (2010) Emergent spaces of reurbanisation: exploring the demographic dimension of inner-city residential change in a European setting. Popul Space Place 16(5):443–463
Halbac-Cotoara-Zamfir R, Egidi G, Salvia R, Salvati L, Sateriano A, Gimenez-Morera A (2021) Recession, local fertility, and urban sustainability: results of a quasi-experiment in Greece, 1991–2018. Sustainability 13(3):1052

Halbac-Cotoara-Zamfir R, Marucci A, Salvia R, Quaranta G, Sateriano A, Cecchini M, Bianchini L (2022) Caring of the Fringe? Mediterranean desertification between peri-urban ecology and socioeconomics. Sustainability 14(3):1426
Hall P (1997) The future of the metropolis and its form. Reg Stud 31:211–220
Ismael HM (2021) Urban form study: the sprawling city—review of methods of studying urban sprawl. GeoJournal 86(4):1785–1796
Jacobs-Crisioni C, Rietveld P, Koomen E (2014) The impact of spatial aggregation on urban development analyses. Appl Geogr 47:46–56
Kaika M (2003) Constructing scarcity and sensationalising water politics: 170 days that shook Athens. Antipode 35(5):919–954
Kallis G (2010) Coevolution in water resource development. The vicious cycle of water supply and demand in Athens, Greece. Ecol Econ 69(4):796–809
Karadimitriou N, Maloutas T, Arapoglou VP (2021) Multiple deprivation and urban development in Athens, Greece: spatial trends and the role of access to housing. Land 10(3):290
Kasanko M, Barredo JI, Lavalle C, McCormick N, Demicheli L, Sagris V, Brezger A (2006) Are European cities becoming dispersed? A comparative analysis of fifteen European urban areas. Landsc Urban Plan 77:111–130
Kazemzadeh-Zow A, Zanganeh Shahraki S, Salvati L, Neisani Samani N (2017) A spatial zoning approach to calibrate and validate urban growth models. Int J Geogr Inf Syst 31(4):763–782
Keddie J, Tonkiss F (2010) The market and the plan: housing, urban renewal and socio-economic change in London. City Cult Soc 1(2):57–67
Kemeny T, Storper M (2012) The sources of urban development: wage, housing, and amenity gaps across American cities. J Reg Sci 52(1):85–108
Koprowska K, Łaszkiewicz E, Kronenberg J (2020) Is urban sprawl linked to green space availability? Ecol Ind 108:105723
Kourtit K, Nijkamp P, Reid N (2014) The new urban world: challenges and policy. Appl Geogr 49:1–3
Kroll F, Kabisch N (2012) The relation of diverging urban growth processes and demographic change along an urban-rural gradient. Popul Space Place 18:260–276
Laidley T (2016) Measuring sprawl: a new index, recent trends, and future research. Urban Affairs Rev 52(1):66–97
Lamonica GR, Recchioni MC, Chelli FM, Salvati L (2020) The efficiency of the cross-entropy method when estimating the technical coefficients of input–output tables. Spat Econ Anal 15(1):62–91
Lampinen J, Tuomi M, Fischer LK, Neuenkamp L, Alday JG, Bucharova A, Klaus VH (2021) Acceptance of near-natural greenspace management relates to ecological and socio-cultural assigned values among European urbanites. Basic Appl Ecol 50:119–131
Le Goix R (2005) Gated communities: sprawl and social segregation in Southern California. Hous Stud 20(2):323–343
Lee S (2011) Metropolitan growth patterns and socio-economic disparity in Six US Metropolitan areas 1970–2000. Int J Urban Reg Res 35(5):988–1011
Legras S, Cavailhès J (2016) Environmental performance of the urban form. Reg Sci Urban Econ 59:1–11
Lindstrom MJ, Bartling H (2003) Suburban sprawl: culture, theory, and politics. Rowman & Littlefield, New York
Liu L, Meng L (2020) Patterns of urban sprawl from a global perspective. J Urban Plann Dev 146(2):04020004
Longhi C, Musolesi A (2007) European cities in the process of economic integration: towards structural convergence. Ann Reg Sci 41:333–351
Lopez R, Hynes HP (2003) Sprawl in the 1990s: measurement, distribution, and trends. Urban Affairs Rev 38(3):325–355
Magliocca N, McConnell V, Walls M, Safirova E (2012) Zoning on the urban fringe: results from a new approach to modeling land and housing markets. Reg Sci Urban Econ 42(1–2):198–210

Makse HA, Andrade JS, Batty M, Havlin S, Stanley HE (1998) Modeling urban growth patterns with correlated percolation. Phys Rev E 58(6):7054
Maloutas T (2004) Segregation and residential mobility. Socially entrapped social mobility and its impact on segregation in Athens. Eur Urban Reg Stud 11(2):171–187
Maloutas T (2007) Socio-economic classification models and contextual difference: the "European socio-economic classes" (ESeC) from a South European angle. South Eur Soc Polit 12(4):443–460
Masini E, Tomao A, Barbati A, Corona P, Serra P, Salvati L (2019) Urban growth, land-use efficiency and local socioeconomic context: a comparative analysis of 417 metropolitan regions in Europe. Environ Manage 63:322–337
Metzger JT (2000) Planned abandonment: The neighborhood life-cycle theory and national urban policy. Hous Policy Debate 11(1):7–40
Morelli GV, Rontos K, Salvati L (2014) Between suburbanisation and re-urbanisation? Revisiting the urban life cycle in a Mediterranean compact city. Urban Res Pract 7(1):74–88
Mouratidis K (2019) Compact city, urban sprawl, and subjective well-being. Cities 92:261–272
Mudu P (2006) Patterns of segregation in contemporary Rome. Urban Geogr 27(5):422–440
Muñoz F (2003) Lock living: urban sprawl in Mediterranean cities. Cities 20(6):381–385
Musterd S (2005) Social and ethnic segregation in Europe: levels, causes and effects. J Urban Aff 27(3):331–348
Musterd S, Marcińczak S, Van Ham M, Tammaru T (2017) Socioeconomic segregation in European capital cities. Increasing separation between poor and rich. Urban Geogr 38(7):1062–1083
Neuman M, Hull A (2009) The futures of the city region. Reg Stud 43(6):777–787
Nickayin SS, Tomao A, Quaranta G, Salvati L, Gimenez Morera A (2020) Going toward resilience? Town planning, peri-urban landscapes, and the expansion of Athens, Greece. Sustainability 12(24):10471
Orenstein DE, Frenkel A, Jahshan F (2014) Methodology matters: measuring urban spatial development using alternative methods. Environ Plann B Plann des 41:3–23
Pacione M (2003) Urban environmental quality and human wellbeing—a social geographical perspective. Landsc Urban Plan 65:19–30
Parés M, March H, Saurí D (2013) Atlantic gardens in Mediterranean climates: understanding the production of suburban natures in Barcelona. Int J Urban Reg Res 37:328–347
Parr J (2014) The regional economy, spatial structure and regional urban systems. Reg Stud 48(12):1926–1938
Partridge MD, Rickman DS, Ali K, Olfert MR (2008) The geographic diversity of U.S. nonmetropolitan growth dynamics: a geographically weighted regression approach. Land Econ 84(2):241–266
Pili S, Grigoriadis E, Carlucci M, Clemente M, Salvati L (2017) Towards sustainable growth? A multi-criteria assessment of (changing) urban forms. Ecol Ind 76:71–80
Power A (2001) Social exclusion and urban sprawl: is the rescue of cities possible? Reg Stud 35(8):731–742
Razin E, Rosentraub M (2000) Are fragmentation and sprawl interlinked? North American Evidence. Urban Affairs Rev 35(6):821–836
Richardson H, Chang-Hee C (2004) Urban sprawl in Western Europe and the United States. Ashgate, London
Roberts S (1991) A critical evaluation of the city life cycle idea. Urban Geogr 12(1):431–449
Rodrigo-Comino J, Egidi G, Sateriano A, Poponi S, Mosconi EM, Gimenez Morera A (2021) Suburban fertility and metropolitan cycles: insights from European cities. Sustainability 13(4):2181
Rontos K, Grigoriadis S, Sateriano A, Syrmali M, Vavouras I, Salvati L (2016) Lost in protest, found in segregation: divided cities in the light of the 2015 'Oki' referendum in Greece. City Cult Soc 7(3):139–148

Rontos K, Papazoglou M, Syrmali ME, Sateriano A (2022) Towards the eastern gate to Europe: factors shaping perceptions and attitudes towards migrants in Lesvos Island (Greece), 2016. Reg Stat 12(1):104–134

Rubiera-Morollón F, Garrido-Yserte R (2020) Recent literature about urban sprawl: a renewed relevance of the phenomenon from the perspective of environmental sustainability. Sustainability 12(16):6551

Sallustio L, Pettenella D, Merlini P, Romano R, Salvati L, Marchetti M, Corona P (2018) Assessing the economic marginality of agricultural lands in Italy to support land use planning. Land Use Policy 76:526–534

Salvati L (2014) Towards a polycentric region? The socioeconomic trajectory of Rome, an 'eternal' Mediterranean City. Tijdschr Econ Soc Geogr 105(3):268–284

Salvati L (2022) Endogenous population dynamics and Metropolitan cycles: long-term evidence from Athens, an eternally Mediterranean City. Eur J Popul 38(5):835–860

Salvati L, Carlucci M (2016) Patterns of sprawl: the socioeconomic and territorial profile of dispersed urban areas in Italy. Reg Stud 50(8):1346–1359

Salvati L, Sateriano A (2015) In-between coast and inland: a long-term assessment of demographic dynamics, agriculture and urban growth in Rome prefecture. Rendiconti Lincei 26:605–614

Salvati L, Serra P (2016) Estimating rapidity of change in complex urban systems: a multidimensional, local-scale approach. Geogr Anal 48(2):132–156

Salvati L, Morelli VG, Rontos K, Sabbi A (2013) Latent exurban development: city expansion along the rural-to-urban gradient in growing and declining regions of southern Europe. Urban Geogr 34(3):376–394

Salvati L, Gargiulo Morelli V (2014) Unveiling urban sprawl in the Mediterranean Region: towards a latent urban transformation? Int J Urban Reg Res 38(6):1935–1953

Salvati L, Ridolfi E, Pujol DS, Ruiz PS (2016) Latent sprawl, divided Mediterranean landscapes: urban growth, swimming pools, and the socio-spatial structure of Athens, Greece. Urban Geogr 37(2):296–312

Salvati L, Sateriano A, Grigoriadis E, Carlucci M (2017) New wine in old bottles: the (changing) socioeconomic attributes of sprawl during building boom and stagnation. Ecol Econ 131:361–372

Salvati L, Ferrara A, Chelli F (2018) Long-term growth and metropolitan spatial structures: an analysis of factors influencing urban patch size under different economic cycles. Geografisk Tidsskrift—Dan J Geogr 118(1):56–71

Salvati L, Ciommi MT, Serra P, Chelli FM (2019) Exploring the spatial structure of housing prices under economic expansion and stagnation: The role of socio-demographic factors in metropolitan Rome, Italy. Land Use Policy 81:143–152

Salvia R, Quaranta V, Sateriano A, Quaranta G (2022) Land resource depletion, regional disparities, and the claim for a renewed 'sustainability thinking' under early desertification conditions. Resources 11(3):28

Sassen S (2010) The city: Its return as a lens for social theory. City Cult Soc 1(1):3–11

Saurí D (2003) Lights and shadows of urban water demand management. The case of the metropolitan region of Barcelona. Eur Plann Stud 11(3):229–243

Schuetze T, Chelleri L, Salvati L (2015) Integrating resilience with urban sustainability in neglected neighborhoods: challenges and opportunities of transitioning to decentralized water management in Mexico city. Habitat Int 48:122–130

Scott AJ (2008) Resurgent metropolis: economy, society and urbanization in an interconnected world. Int J Urban Reg Res 32(3):548–564

Scott AJ, Storper M (2015) The nature of cities: the scope and limits of urban theory. Int J Urban Reg Res 39(1):1–15

Scott AJ, Carter C, Reed MR, Stonyer B, Coles R (2013) Disintegrated development at the rural-urban fringe: re-connecting spatial planning theory and practice. Prog Plan 83:1–52

Seifollahi-Aghmiuni S, Kalantari Z, Egidi G, Gaburova L, Salvati L (2022) Urbanisation-driven land degradation and socioeconomic challenges in peri-urban areas: insights from Southern Europe. Ambio 51(6):1446–1458

Serra P, Vera A, Tulla AF, Salvati L (2014) Beyond urban–rural dichotomy: exploring socioeconomic and land-use processes of change in Spain (1991–2011). Appl Geogr 55:71–81

Sinclair R (1967) Von Thunen and urban sprawl. Ann Assoc Am Geogr 57:72–87

Smiraglia D, Ceccarelli T, Bajocco S, Perini L, Salvati L (2015) Unraveling landscape complexity: land use/land cover changes and landscape pattern dynamics (1954–2008) in contrasting peri-urban and agro-forest regions of northern Italy. Environ Manage 56:916–932

Storper M, Scott AJ (2016) Current debate in urban theory: a critical assessment. Urban Stud 53(6):1114–1136

Swyngedouw E (2004) Social power and the urbanization of water. Oxford University Press, Oxford, UK

Tomao A, Quatrini V, Corona P, Ferrara A, Lafortezza R, Salvati L (2017) Resilient landscapes in Mediterranean urban areas: understanding factors influencing forest trends. Environ Res 156:1–9

Tomao A, Quaranta G, Salvia R, Vinci S, Salvati L (2021) Revisiting the 'southern mood'? Post-crisis Mediterranean urbanities between economic downturns and land-use change. Land Use Policy 111:105740

Townsend A (2009) Integration of economic and spatial planning across scales. Int J Public Sect Manag 22(7):643–659

Tsilimigkas G, Kizos T (2014) Space, pressures and the management of the Greek landscape. Geografiska Annaler Ser B Human Geogr 96(2):159–175

Vicino TJ, Hanlon B, Short JR (2007) Megalopolis 50 years on: the transformation of a city region. Int J Urban Reg Res 31(2):344–367

Vidal M, Domene E, Saurí D (2011) Changing geographies of water-related consumption: residential swimming pools in suburban Barcelona. Area 43(1):67–75

Vinci S, Egidi G, López Gay A, Salvati L (2022) Population growth and urban management in metropolitan regions: the contribution of natural balance and migration to polycentric development in Barcelona. Appl Spat Anal Policy 15(1):71–94

Vinci S, Vardopoulos I, Salvati L (2023) A tale of a shrinking city? Exploring the complex interplay of socio-demographic dynamics in the recent development of Attica, Greece. Cities 132:104089

Wilson GA, Kelly CL, Briassoulis H, Ferrara A, Quaranta G, Salvia R, Detsis V, Curfs M, Cerda A, El-Aich A, Liu H, Kosmas C, Alados CL, Imeson A, Landgrebe-Trinkunaite R, Salvati L, Naumann S, Danwen H, Iosifides T, Kizos T, Mancino G, Nolè A, Jiang M, Zhang P (2017) Social memory and the resilience of communities affected by land degradation. Land Degrad Dev 28(2):383–400

Zambon I, Serra P, Sauri D, Carlucci M, Salvati L (2017) Beyond the 'Mediterranean city': Socioeconomic disparities and urban sprawl in three Southern European cities. Geografiska Annaler Ser B Human Geogr 99(3):319–337

Zambon I, Colantoni A, Salvati L (2019) Horizontal vs vertical growth: understanding latent patterns of urban expansion in large metropolitan regions. Sci Total Environ 654:778–785

Zwiers M, Bolt G, van Ham M, Van Kempen R (2016) The global financial crisis and neighbourhood decline. Urban Geogr 37(5):664–684

Chapter 3
Regional Development, Spatial Disparities, and Urbanization: A Comparative Approach

Abstract Urban concentration and economic agglomeration have played an important role in development processes along the last century and were associated with population growth in recent times, suggesting the rising importance of local (non-economic) drivers. The largest metropolitan regions, however, seem to deviate from this path and adhere to more individual expansion models. Territorial characteristics may also influence such paths, suggesting how metropolitan growth is broadly heterogeneous and almost unpredictable. Understanding the intimate mechanisms of metropolitan growth means to seriously consider some active constraints to urban expansion such as the available land to building, the peculiarity of some local contexts, demographic frictions (e.g. because of aging), and regional planning. When regions plan to mitigate ecological and socioeconomic risk for resident populations, sustainability and resilience targets should be regarded as intrinsically associated with metropolitan expansion. Based on these premises, the present chapter provides a brief review of theoretical approaches and practical applications alimenting a 'resilience debate' with a specific focus on urban affairs. The ability to adapt to economic/technological change—affecting evolutionary dynamics of regional economies—was intended as a primary key to face with sociopolitical challenges and the demanding request for sustainable development and environmental quality preservation in metropolitan regions.

Keywords Economic growth · Spatial planning · Territorial divides · Indicators · Europe

3.1 Introduction

Important transformations in land use have characterized the long-term development of Mediterranean countries since World War II, influencing natural and agricultural ecosystems, and slowly modifying the typical traits of extensiveness and environmental compatibility that characterized rural landscapes in most regions of Southern Europe (Sun et al. 2020). Currently, it was argued how agricultural landscapes are

L. Salvati, *Economic Growth, Regional Disparities, and Urban Sprawl*, Springer Geography, https://doi.org/10.1007/978-3-031-85799-7_3

evolving more rapidly than in the recent past (Khan et al. 2021). Coastal urbanization and recurrent industrialization of inland districts, infrastructural growth, and exurban development, are the result of an inexorable drift of regional economic systems towards advanced tertiary sectors (Angel et al. 2011). Changes over time in consumer preferences and lifestyles—in both urban and rural areas—have given a significant input to the inherent transformation of rural territories (e.g. Davis 1955).

In recent years, production processes moving from central locations to suburban districts had significant consequences on local development paths, outlining the increased importance of 'total governance', or the substantial involvement of territorial actors. Borrowing some of the most frequent policy recommendations from the Organization for Economic Cooperation and Development (European Commission 1999), a development strategy can be identified in such contexts, that is based on the extrinsic growth of business clusters to increase productivity through shared technologies (Murtin 2013). This strategy is grounded on distinctive leverages, including:

(i) the full involvement of (global and local) finance for operational recovery of fringe rural spaces,
(ii) an enhanced relationship between education centers, universities, and the entrepreneurial system promoting widespread innovation, and
(iii) a closer relationship between urban areas and the system of historical (rural) settlements and municipalities with high cultural and artistic value.

At the same time, and especially from the 1970s onwards, agriculture has progressively lost its economic role in the most accessible districts surrounding cities—likely with the only exception of Eastern (socialist) countries (Profiroiu et al. 2020). Mechanization, irrigation, soil compaction, and water regulation with great capital impacts, have pushed towards a latent process of crop intensification which, still today, is reflected in poorly diversified landscapes typical of fringe districts, featuring a significant scarcity of natural habitats, biodiversity traits, and high-quality visual elements (Shaw et al. 2020). At the same time, the economic decline of agriculture has involved not only marginal, mountainous, or poorly accessible areas (Gao et al. 2023), but also more accessible districts oriented toward industrial production. These have experienced a progressive 'land speculation' leveraging conversion to urban use (Herrero-Jauregui and Conception 2023). These processes basically led to low-density suburbanization, consolidating a land-use mix that formed an entropic and often undefined peri-urban matrix of non-urban (and non-rural) landscapes (Chen and Chi 2022). For such reasons, an integrated vision 'from both sides of the fringe' is increasingly needed when reading and interpreting metropolitan complexity in advanced economies (Beckers et al. 2020).

3.2 Compact Versus Dispersed: The Evolving Paradigm of Mediterranean Urbanization

The paradigm of 'compact cities' was clearly defined from a land-use perspective, and its implications were outlined in the recent literature, considering together environmental, social and economic benefits and/or negative impacts (Tian et al. 2022). Compact settlement models were classically regarded as land-saving ways of (re)structuring metropolitan regions and regulating urban growth in a mostly radiocentric development path (Glaeser 2020). Although they may occasionally produce negative externalities (e.g. pollution, congestion, scarcity of green spaces) and impact landscapes, especially within the strict boundary of urban areas (Hamnett 2020), these paths reflect a developmental model with low land take rates (Fig. 3.1).

In compact models, the explicit hierarchy from high-density metropolitan centers to low-density peripheral villages and isolated settlements (Oddo and Zanini 2022), shaped the multiple relationships between urban areas and neighboring (rural) districts, encompassing socioeconomic and ecological dimensions (Mukhopadhyay et al. 2020). The importance of urban–rural gradients is reflected in a clear land-use segregation and distinctive socioeconomic roles for central and peripheral territories (Henderson and Turner 2020). More specifically, urban areas were intended as consumption places and the location of advanced productions—being intrinsically characterized by congestion and air pollution because of human density (Zhang et al. 2022). Rural areas were, in turn, intended as places specialized in agricultural production, and guarantee the conservation (or reproduction) of natural amenities (Cai et al. 2020), in a context of low (or medium–low) human density.

Fig. 3.1 Traditional vineyards in peri-urban 'Castelli Romani' district, South-East of Rome. *Source* Author's photographic archive

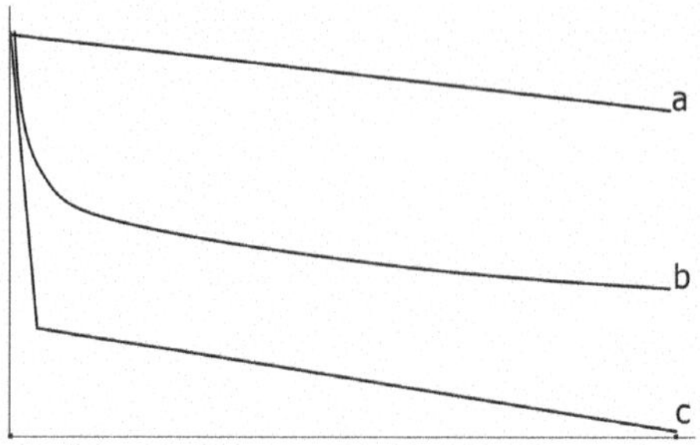

Fig. 3.2 Different metropolitan gradients under fully dispersed (**a**), moderately dense and basically polycentric (**b**) and compact mono-centric (**c**) regions, relating population density (*y*-axis) with the distance from inner cities (*x*-axis). *Source* Author's elaboration

The 'dispersed city' model delineates the inherent dynamics of urban landscapes invading rural countryside through a disorganized, heterogeneous, and disarticulated spillover of built-up settlements—losing a clear 'zoning' of residential, commercial, and industrial land use (Andreotti et al. 2013; Xiong et al. 2020; Pandey et al. 2022). The transition from compact cities towards dispersed metropolises results in the increasing land-use mix along density gradients, which became progressively less steep (Fig. 3.2). In other words, the economic depolarization of inner cities, and fringe sprawl, tended to re-organize rural territories within undefined landscapes (Adger et al. 2020).

Land fragmentation is thus one of the most appropriate keys interpreting human complexity (and, possibly, congestion) of such landscapes, because it defines and quantifies a sort of 'structural disorder' in their spatial configuration (Fig. 3.3). The agricultural traits of these landscapes are (irreversibly, or hardly reversibly) lost (Kwilinski et al. 2023). Natural features of city regions, especially woodland, are damaged or declining in size or ecological functions (Wang et al. 2022). The inherent amplification of the ecological issues typical of urban areas—together with relevant social issues in rural areas—is an evident externality of dispersed growth, leveraging a downward environmental spiral hard to assess and mitigate effectively (Randolph and Storper 2023).

The evolving role of rural areas from a primary production agent (providing agriculture, livestock, and forestry functions and preserving natural amenities) to a mere 'object' serving cities (e.g. as a source of buildable land), was progressively more evident, and outlined in the mainstream literature (Jing et al. 2020). In this perspective, we question if rural areas will inexorably transform into a sort of 'garbage can' of the metropolis (Castells-Quintana and Wenban-Smith 2020). Candidate proofs of this transition are as follows:

(i) industrial and commercial areas are expelled from city centers and re-located within rural spaces (Allen et al. 2004);

Fig. 3.3 Historical stratification of compact settlements in downtown Seville, looking at the Southern countryside beyond the Guadalquivir bridge. *Source* Author's photographic archive

(ii) residential districts reproduce themselves in the same (rural) space via 'obsessively standard' forms, mimicking the density characteristic of the settlement types from which they conceptually derive (Alexandri 2015);
(iii) sprawled low-density villages grow everywhere without a clear planning, lacking an unified 'urban vision' (Fig. 3.4).

These are all elements of a common picture depicting landscapes within contemporary peri-urban regions (Wang et al. 2020). Road infrastructures (and at a less pace, railways) expanded anywhere with the aim at connecting residential and productive settlements originally conceived as functionally distinctive and spatially separated (Solly et al. 2020; Giacalone et al. 2023; Kockelkorn et al. 2023).

Globalization has accelerated the loss of the original functions characteristic of rural areas, involving all the (economic) productions, sectors, and (spatial) scales (Menashe-Oren and Bocquier 2021). 'Globalized' citizens of the diffused metropolis do not feel the preservation of the surrounding agricultural base as a crucial planning and policy issue (Cividino et al. 2020), since primary goods are provided to the expanding urban markets from very far lands, better equipped for large-scale production (Dombi 2021). The same citizens may judge as superfluous to protect the remaining (unsealed) rural territory which produces natural amenities and tourism leisure (Berisha and Cotella 2021), since the increasing human mobility stimulates tourism fluxes moving outside the boundary of the metropolitan region where they

Fig. 3.4 A rapidly changing countryside specialized in olive production and wild livestock into a tourism attraction: Elafonisi island, Lakonia, Southern Peloponnese. *Source* Author's photographic archive

settle (Sahana et al. 2023). At the same time, big pipelines aliment cities with the (supposedly unlimited) quantity of water produced very far from its surroundings (Kaika 2003).

In Mediterranean regions, rural spaces are thus reflecting a sort of bad copy of the surrounding urban landscapes. They represent 'originally rural' landscapes evolving toward urbanities where not only affluent people built-up their villas but where also *bourgeoises* and even economically disadvantaged social classes escape the congestion of inner cities searching for any sort of 'deconcentration' amenities (Zhang 2008). Second homes were transformed to allow a rising urban population not only to live there during the long summer holidays, but instead to pass there a large part of both free and working time all over the year (Richtie and Roser 2024). By escaping the metropolises' congestion, however, urban-natives—now converted into *naive* rural-seekers—create another type of congestion, which reflects the loss of the (hierarchical) structure of rural landscapes surrounding the traditional compact cities (Lowry 1990).

The dispersed second-home settlements (with all ranges of construction types, from the poorest 'villages' to the richest detached villas) are only one of the basic proofs of 'landscape congestion' (Fig. 3.5). They not only dominate coastal landscapes, but also invade the internal lowlands, originally specialized in wheat, fruit trees, and grapevine production (Biasi et al. 2015). They also approach, and even climb, the sloping uplands—original kingdom of olives and old forests in Southern Europe (Bianchini et al. 2021a, b). These dynamics anticipate the uneven environmental (and socioeconomic) issues derived from a weak management of dispersed urban settlements (Bueno-Suarez and Coq-Huelva 2020), within a landscape that has progressively lost its rural traits and urges dedicated policy strategies well beyond standard planning tools (Cazcarro et al., 2012). Considering preservation of peri-urban biodiversity and local traditions at the fringe of mega-city regions as a basic target of United Nations 2030 Sustainability Agenda, the role of Mediterranean landscapes, e.g. those dominated by olive groves, vineyards and a diversified matrix of shrubland, pastures and fallow land, was crucial in settlement containment and fringe sustainability (Colantoni et al. 2017). For instance, having a great cultural, culinary, and aesthetic importance, olive groves characterized Mediterranean peri-urban landscapes in a distinctive way (Cuadrado-Ciuraneta et al. 2017).

Fig. 3.5 An example of 'congested landscapes': a shot of South Valencia agricultural plan from Cullera, with natural maquis vegetation, solar plants, traditional dry cropland, and second-home settlements along the Mediterranean sea coast; author's photographic archive

3.3 'Fast' Urbanization and the 'Slow' Evolution of Rural Landscapes

A comprehensive understanding of (apparent and latent) mechanisms underlying metropolitan growth, provides a more general contribution to socioeconomic development processes at both local and regional scales (Halbac-Cotoara-Zamfir et al. 2021, 2022). The most appropriate analytical strategy should place territories at the very center of the debate on sustainable, spatially balanced, socially cohesive, and environmentally friendly urbanization models (Garcia-Lopez 2019). Within this framework, cities are no longer assumed as places of tradition and history, while becoming regions with increasing complexity in all the pillars of sustainability (Keddie and Tonkiss 2010). Rapidly expanding cities have been seen as internally articulated and fragmented units, difficult to manage, plan, and regulate in their internal growth, with a dominant position with respect to the surrounding territories, also thanks to the continuous extraction of natural resources from the nearest rural districts (Galster et al. 2001; Grout et al. 2016; Ismael 2021). The most recent international reports on urbanization and environmental quality outlined the crucial role of cities in the increasingly pervasive global transformations observed on our planet—from climate change to water scarcity, from poverty to environmental migration (James 2014; Feng et al. 2019; Egidi et al. 2020). This was reflected in the complex demographic trends observed in Europe and in the surrounding areas, as delineated in the comparison of long-term urban (Table 3.1) and rural (Table 3.2) dynamics.

The inherent difficulty in analyzing how metropolitan systems are structured, and consolidate over time, highlights the complexity of this issue, as clearly observed in several policy reports at the continental scale since the mid-1990s (European Commission 1999). Research tools should, therefore, adhere to a multidisciplinary vision that integrates different methodologies and empirical approaches (e.g. Karadimitriou et al. 2021). Quantitative analysis, exploiting the power of the available 'big data', allows to draw increasingly updated maps of urban growth at the individual city level (Buzar et al. 2007; Garcia-Lopez and Muniz 2010; Brueggmann 2019). Implementing a retrospective analysis that covers sufficiently long time periods will allow for characterizing the different socioeconomic dynamics that have shaped metropolitan evolution in the last century (Benassi and Salvati 2020). Based on a large collection of indicators, quantitative analysis of statistical indicators allows for the investigation of integrated dynamics of economic growth and metropolitan development at different observation scales (Balk et al. 2006), outlining common paths among cities and peculiar individual behaviors, distinguishing urban from rural contexts and providing some key regional and country comparisons, of interest when exploring relatively long time intervals.

These paths integrate demographic, social and economic dynamics, and can be traced back to specific historical phases, regional social contexts, and economic cycles, thus contributing to a refined understanding of how metropolitan development and economic growth adhere to long-term sustainability principles (Balsa-Barreiro et al. 2022). In this perspective, some theoretical and empirical reflections about the

Table 3.1 Average annual rate of change in the percent share of urban population in total population of selected countries, 1950–2025

Region, Country	1950–1955	1960–1965	1970–1975	1980–1985	1990–1995	2000–2005	2010–2015	2020–2025
WORLD	**1.33**	**1.06**	**0.61**	**0.92**	**0.83**	**1.04**	**0.86**	**0.75**
Northern Africa	*1.99*	*1.85*	*1.24*	*1.05*	*0.69*	*0.42*	*0.37*	*0.49*
Algeria	3.26	4.20	0.42	1.94	1.45	1.26	0.96	0.66
Egypt	1.74	0.94	0.86	0.04	– 0.31	0.11	– 0.11	0.31
Libya	3.43	4.18	4.69	1.47	0.07	0.18	0.31	0.37
Morocco	1.15	1.64	1.79	1.66	1.32	0.68	0.94	0.81
Sudan	4.59	4.37	2.73	2.78	2.38	0.16	0.48	1.07
Tunisia	1.75	1.15	1.80	1.25	1.18	0.56	0.42	0.47
Western Sahara	0.06	2.97	6.91	1.67	0.34	0.06	0.06	0.10
Western Asia	*2.40*	*2.07*	*1.78*	*1.95*	*0.42*	*0.61*	*0.61*	*0.41*
Armenia	2.61	1.63	1.21	0.31	– 0.41	– 0.23	– 0.11	0.26
Azerbaijan	1.81	– 0.58	0.74	0.27	– 0.58	0.39	0.48	0.71
Bahrain	3.19	0.04	0.28	0.24	0.06	0.01	0.08	0.14
Cyprus	2.61	1.39	2.97	1.97	0.37	– 0.11	– 0.18	0.11
Georgia	1.64	1.11	0.95	0.64	– 0.44	0.36	0.68	0.71
Iraq	1.38	3.32	1.78	0.97	– 0.27	0.08	0.24	0.34
Israel	0.00	1.04	0.57	0.28	0.11	0.07	0.08	0.10
Jordan	3.11	1.29	0.59	2.20	1.30	0.31	0.95	0.21
Kuwait	2.40	1.14	0.85	0.65	0.02	0.20	0.00	0.00
Lebanon	2.52	3.69	2.38	1.50	0.40	0.15	0.18	0.19
Oman	6.60	6.18	5.06	3.65	1.62	0.23	1.58	0.74
Qatar	0.61	0.41	0.13	0.16	0.47	0.22	0.09	0.04

(continued)

Table 3.1 (continued)

Region, Country	1950–1955	1960–1965	1970–1975	1980–1985	1990–1995	2000–2005	2010–2015	2020–2025
Saudi Arabia	3.96	4.31	3.63	1.96	0.54	0.28	0.27	0.26
State of Palestine	1.71	2.09	1.86	0.85	0.73	0.30	0.33	0.38
Syrian Arab Republic	1.20	1.67	0.78	0.50	0.47	0.69	– 1.27	1.12
Turkey	2.84	1.65	1.68	3.61	0.96	0.94	0.77	0.57
United Arab Emirates	3.39	1.12	0.00	– 0.23	– 0.19	0.50	0.37	0.27
Yemen	4.54	3.84	2.20	2.11	2.54	1.94	1.80	1.63
EUROPE	**1.05**	**1.01**	**0.74**	**0.38**	**0.16**	**0.25**	**0.25**	**0.32**
Western Europe	*0.59*	*0.56*	*0.24*	*0.05*	*0.25*	*0.35*	*0.21*	*0.24*
Eastern Europe	*2.22*	*1.57*	*1.29*	*0.74*	*0.05*	*0.08*	*0.11*	*0.28*
Northern Europe	*0.25*	*0.31*	*0.36*	*0.11*	*0.11*	*0.26*	*0.33*	*0.29*
Southern Europe	*1.07*	*1.30*	*0.81*	*0.32*	*0.29*	*0.40*	*0.41*	*0.44*
Albania	5.38	0.34	0.60	0.77	1.32	2.26	1.93	1.25
Andorra	4.49	3.73	1.70	0.66	– 0.22	– 0.46	– 0.11	– 0.04
Bosnia/Herzegovina	3.37	3.62	2.80	1.12	0.78	0.73	0.70	0.85
Croatia	3.10	2.97	1.75	0.83	0.48	0.33	0.36	0.61
Greece	0.73	1.39	0.83	0.34	0.18	0.47	0.45	0.39
Italy	0.99	0.82	0.42	0.06	0.06	0.15	0.36	0.44
Malta	0.22	– 0.07	0.01	0.01	0.12	0.27	0.07	0.07
Montenegro	3.97	3.66	3.27	2.79	2.10	1.30	0.51	0.49
Portugal	1.10	1.06	0.99	1.14	1.29	1.12	0.95	0.78
San Marino	1.27	2.13	3.57	1.67	0.28	0.22	0.21	0.11
Serbia	3.69	3.26	1.63	0.93	0.49	0.42	0.25	0.42

(continued)

Table 3.1 (continued)

Region, Country	1950–1955	1960–1965	1970–1975	1980–1985	1990–1995	2000–2005	2010–2015	2020–2025
Slovenia	3.57	2.83	2.71	0.63	0.09	0.30	0.42	0.60
Spain	0.87	1.60	1.04	0.39	0.13	0.26	0.29	0.31
TFYR N. Macedonia	3.87	3.42	1.43	0.82	0.61	− 0.35	0.11	0.61

Source Author's elaboration on World Urbanization Prospects, United Nations Statistical Office, 2018

Table 3.2 Average annual rate of change in the percent share of rural population in total population of selected countries, 1950–2025

Region, Country	1950–1955	1960–1965	1970–1975	1980–1985	1990–1995	2000–2005	2010–2015	2020–2025
WORLD	**– 0.58**	**– 0.56**	**– 0.36**	**– 0.62**	**– 0.65**	**– 0.95**	**– 0.96**	**– 1.01**
Northern Africa	*– 0.75*	*– 0.91*	*– 0.76*	*– 0.77*	*– 0.60*	*– 0.40*	*– 0.38*	*– 0.56*
Algeria	– 1.04	– 2.17	– 0.28	– 1.63	– 1.70	– 2.05	– 2.15	– 1.98
Egypt	– 0.87	– 0.60	– 0.63	– 0.03	0.23	– 0.08	0.08	– 0.23
Libya	– 0.93	– 1.83	– 6.04	– 3.95	– 0.22	– 0.60	– 1.14	– 1.63
Morocco	– 0.43	– 0.72	– 1.01	– 1.25	– 1.32	– 0.80	– 1.38	– 1.50
Sudan	– 0.38	– 0.60	– 0.59	– 0.76	– 1.04	– 0.08	– 0.24	– 0.61
Tunisia	– 0.89	– 0.72	– 1.51	– 1.37	– 1.75	– 1.01	– 0.86	– 1.12
Western Sahara	– 0.03	– 1.51	– 7.13	– 6.33	– 1.89	– 0.35	– 0.36	– 0.69
Western Asia	*– 1.07*	*– 1.29*	*– 1.57*	*– 2.36*	*– 0.68*	*– 1.13*	*– 1.39*	*– 1.10*
Armenia	– 1.98	– 1.87	– 1.96	– 0.62	0.82	0.41	0.19	– 0.46
Azerbaijan	– 1.66	0.62	– 0.77	– 0.31	0.66	– 0.42	– 0.57	– 0.96
Bahrain	– 7.50	– 0.20	– 1.54	– 1.58	– 0.42	– 0.09	– 0.65	– 1.23
Cyprus	– 1.13	– 0.81	– 2.33	– 3.17	– 0.77	0.24	0.37	– 0.23
Georgia	– 1.03	– 0.89	– 0.92	– 0.73	0.53	– 0.41	– 0.88	– 1.09
Iraq	– 0.79	– 2.92	– 2.54	– 1.98	0.60	– 0.17	– 0.54	– 0.86
Israel	0.00	– 3.89	– 3.37	– 2.31	– 1.08	– 0.73	– 0.88	– 1.26
Jordan	– 2.08	– 1.43	– 0.78	– 3.83	– 4.07	– 1.15	– 7.12	– 2.39
Kuwait	– 4.56	– 3.83	– 6.01	– 18.2	– 1.24	0.00	0.00	0.00
Lebanon	– 1.31	– 3.23	– 4.09	– 4.93	– 2.12	– 0.92	– 1.26	– 1.60
Oman	– 0.75	– 1.47	– 2.59	– 4.01	– 3.59	– 0.59	– 5.73	– 5.45
Qatar	– 2.76	– 2.58	– 0.98	– 1.44	– 7.32	– 6.97	– 7.03	– 5.43

(continued)

Table 3.2 (continued)

Region, Country	1950–1955	1960–1965	1970–1975	1980–1985	1990–1995	2000–2005	2010–2015	2020–2025
Saudi Arabia	– 1.22	– 2.32	– 4.18	– 4.43	– 1.87	– 1.16	– 1.26	– 1.46
State of Palestine	– 1.09	– 1.81	– 2.47	– 1.49	– 1.62	– 0.79	– 0.97	– 1.29
Syrian Arab Republic	– 0.61	– 1.04	– 0.62	– 0.45	– 0.46	– 0.78	1.49	– 1.49
Turkey	– 1.03	– 0.81	– 1.12	– 3.35	– 1.49	– 1.84	– 2.01	– 1.94
United Arab Emirates	– 5.00	– 3.47	0.00	0.92	0.69	– 2.16	– 2.10	– 1.91
Yemen	– 0.32	– 0.43	– 0.36	– 0.45	– 0.73	– 0.74	– 0.90	– 1.06
EUROPE	**– 1.18**	**– 1.44**	**– 1.34**	**– 0.81**	**– 0.39**	**– 0.63**	**– 0.69**	**– 0.99**
Western Europe	*– 1.11*	*– 1.29*	*– 0.63*	*– 0.14*	*– 0.73*	*– 1.15*	*– 0.79*	*– 0.98*
Eastern Europe	*– 1.61*	*– 1.62*	*– 1.81*	*– 1.38*	*– 0.11*	*– 0.17*	*– 0.25*	*– 0.67*
Northern Europe	*– 0.58*	*– 0.81*	*– 1.05*	*– 0.36*	*– 0.38*	*– 0.93*	*– 1.40*	*– 1.43*
Southern Europe	*– 0.97*	*– 1.47*	*– 1.19*	*– 0.55*	*– 0.54*	*– 0.81*	*– 0.95*	*– 1.20*
Albania	– 1.66	– 0.15	– 0.29	– 0.40	– 0.80	– 1.79	– 2.33	– 2.24
Andorra	– 3.48	– 6.81	– 8.85	– 9.89	3.63	4.90	0.83	0.30
Bosnia/Herzegovina	– 0.59	– 0.96	– 1.16	– 0.65	– 0.52	– 0.56	– 0.60	– 0.86
Croatia	– 0.99	– 1.43	– 1.27	– 0.77	– 0.51	– 0.38	– 0.45	– 0.85
Greece	– 0.82	– 1.92	– 1.58	– 0.78	– 0.46	– 1.31	– 1.54	– 1.62
Italy	– 1.24	– 1.26	– 0.78	– 0.11	– 0.12	– 0.32	– 0.80	– 1.12
Malta	– 1.84	0.61	– 0.08	– 0.08	– 1.21	– 3.66	– 1.19	– 1.29
Montenegro	– 0.65	– 0.95	– 1.35	– 1.83	– 2.16	– 1.99	– 0.95	– 1.07
Portugal	– 0.52	– 0.60	– 0.66	– 0.90	– 1.27	– 1.42	– 1.55	– 1.62
San Marino	– 1.02	– 2.28	– 6.87	– 9.44	– 2.82	– 3.35	– 5.34	– 4.68
Serbia	– 1.06	– 1.50	– 1.16	– 0.83	– 0.51	– 0.48	– 0.31	– 0.56

(continued)

Table 3.2 (continued)

Region, Country	1950–1955	1960–1965	1970–1975	1980–1985	1990–1995	2000–2005	2010–2015	2020–2025
Slovenia	− 1.00	− 1.23	− 1.78	− 0.60	− 0.09	− 0.32	− 0.48	− 0.76
Spain	− 0.98	− 2.30	− 2.20	− 1.08	− 0.41	− 0.86	− 1.11	− 1.34
TFYR N. Macedonia	− 1.35	− 2.02	− 1.36	− 0.98	− 0.87	0.48	− 0.15	− 0.89

Source Author's elaboration on World Urbanization Prospects, United Nations Statistical Office, 2018

role urbanization processes and demographic dynamics play in any given sustainable development path, should be proposed, considering together development needs and environmental constraints (Domene and Sauri 2006; Chorianopoulos et al. 2010; Benassi et al. 2023). Empirical data are referred to the old continent and, particularly, to Mediterranean Europe, a paradigmatic region of extreme interest for sustainability studies, given its long-term growth path as far as demographic increase, urban structure, and socioeconomic fragility are concerned (Dura-Guimera 2003; Cassiers and Kesteloot 2012; Di Feliciantonio and Salvati 2015).

3.4 Urbanization as the Engine of Economic Growth

(i) The joint analysis of the results of multiple exploratory approaches on empirical data and statistics enables to derive an interpretative model of the 'life cycle' of a given metropolitan region, accounting for sequential economic stages that reflect distinctive local contexts (Couch et al. 2007). In Europe, as suggested in Fig. 3.6, there is a broad consensus on the end of a specific (and possibly long) cycle, while approaching a new urban experience, that is considered rather distinct from previous ones (Benassi et al. 2020). This interpretation prefigures a completely new development path well characterized over time (i.e. distinguishing change factors active on medium-short and longer scales), space (i.e. identifying change factors more active on local, regional, and wider spatial scales), and economic specialization. Despite the wealth of information, future urban (and suburban) models in the old continent require additional, comparative, investigation. For instance, a combined study of the progressive decline of selected metropolitan regions, may delineate the consolidation of less sustainable development paths (Arbaci 2008; Arapoglou and Sayas 2009; De Muro et al. 2011), possibly associated with an increased consumption of land and natural resources because of settlement expansion (Domene et al. 2005).

At the same time, the factors behind urban density became less associated with the net rate of urban growth, documenting the rising importance of other (non-economic) contextual forces (e.g. Arbaci and Malheiros 2010), and the emergence of individualistic growth paths (Fig. 3.7). The knowledge acquired so far will contribute to derive a shared framework on metropolitan growth, taking account of some constraints—such as land availability and spatial planning (Iosifides and Politides 2005; Koprowska et al. 2020). With this perspective in mind, regional growth processes, and the translation into fine-grained urban expansion, should be regarded as a powerful expression of cultural, intellectual, political, and organization traits of any human society (Kemeny and Storper 2012).

In this view, any strategy implementing quality of life as a basic aspect of sustainable development in metropolitan regions (sensu Mouratidis 2019) may reflect the role of a composite social-economic-environmental dimension of growth and change (Liu and Meng 2020). The empirical findings of these studies will support specific

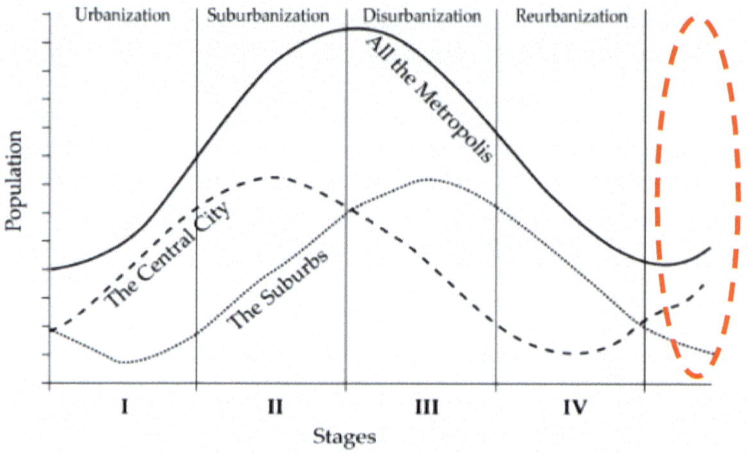

Fig. 3.6 A schematic representation of the life cycle of cities, distinguishing four development waves in metropolitan regions of Southern Europe and indicating the present condition of spatial heterogeneity and economic volatility (dashed vertical ellipse). *Source* Author's elaboration of original materials derived from Kirk (1996) and Roberts (1991)

Fig. 3.7 Spatial distribution and areal size of the metropolitan regions hosting capital cities in the old continent. *Source* Author's elaboration of Eurostat official statistics and spatial geo-databases

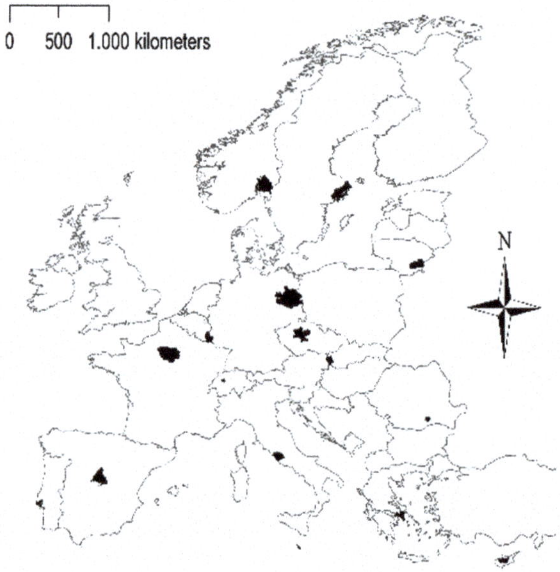

policies—e.g. those envisaged in United Nations Agenda 2030 Sustainable Development Goals (see Fig. 3.8)—targeting urban containment and a balanced growth of settlements in both wealthy and economically depressed regions (Maloutas 2004).

11.1 - Ensure access for all to adequate, safe and affordable housing and basic services and upgrade slums.
11.2 - Provide access to safe, affordable, accessible and sustainable transport systems for all, improving road safety, notably by expanding public transport, with special attention to the needs of those in vulnerable situations, women, children, persons with disabilities and older persons.
11.3 - Enhance inclusive and sustainable urbanization and capacity for participatory, integrated and sustainable human settlement planning and management in all countries.
11.4 - Strengthen efforts to protect and safeguard the world's cultural and natural heritage.
11.5 - Significantly reduce the number of deaths and the number of people affected and substantially decrease the direct economic losses relative to global gross domestic product caused by disasters, including water-related disasters, with a focus on protecting the poor and people in vulnerable situations.
11.6 - Reduce the adverse per capita environmental impact of cities, including by paying special attention to air quality and municipal and other waste management
11.7 - Provide universal access to safe, inclusive and accessible, green and public spaces, in particular for women and children, older persons and persons with disabilities.
11.A - Support positive economic, social and environmental links between urban, peri-urban and rural areas by strengthening national and regional development planning.
11.B - Increase the number of cities and human settlements adopting and implementing integrated policies and plans towards inclusion, resource efficiency, mitigation and adaptation to climate change, resilience to disasters, and develop and implement, in line with the Sendai Framework for Disaster Risk Reduction 2015-2030, holistic disaster risk management at all levels.
11.C - Support least developed countries, including through financial and technical assistance, in building sustainable and resilient buildings utilizing local materials.

Fig. 3.8 Specific targets of Sustainable Development Goal 11 dealing with cities. *Source* United Nations Agenda 2030

3.5 Settlement Models and Regional Demographic Change

Since a long time, settlement models and population dynamics are seen as interconnected issues with evident and practical implications for sustainable land management (Orenstein et al. 2014), since a balanced settlement model across various demographic and social dimensions may support urban sustainability (Munoz 2003). A permanent assessment of these factors may, in turn, allow for the implementation of different development scenarios (Masini et al. 2019). Following a double demographic crisis (economic recession first, and Covid-19 pandemic later) that has caused latent paths of urban decline in Southern Europe (Fig. 3.9), analyzing processes of change in regional spatial structures is a key issue for a refined identification of settlement models, distinguishing truly mono-centric from largely polycentric, or mixed and more scattered organizations (Kroll and Kabisch 2012).

Recent literature has documented how the shift from mono-centric spatial organizations to more dispersed structures has determined an increase of morphological entropy and fractal dimension of land parcels, with a declining importance of the

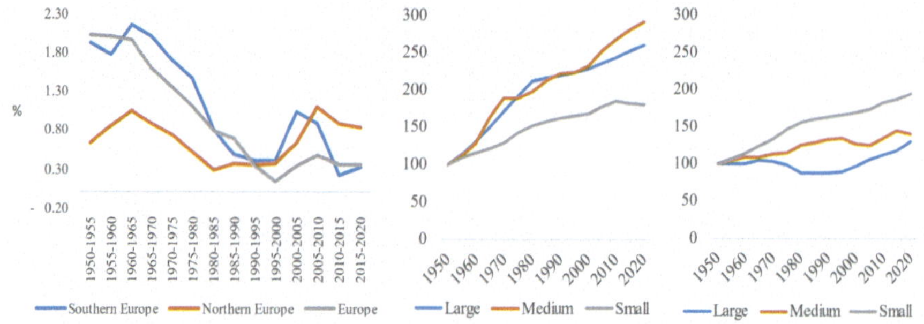

Fig. 3.9 Average annual rate of change of total population (left) and urban population, 1950–2020 (percent); population in cities by size class of urban settlement, 1950–2020 (index number to a fixed base; 1950 = 100) in Southern Europe (middle) and Northern Europe (right). *Source* Author's elaboration on United Nations, Department of Economic and Social Affairs, Population Division (2018). World Urbanization Prospects: The 2018 Revision

linear distance from downtown as a factor of urbanization (Musterd 2005; Mudu 2006; Nickayin et al. 2020). Changes in urban population thus became a key indicator for understanding settlement models of a given socioeconomic system especially when considered together with changes over time in total population (Marti-Hennberg 2005; Maloutas 2007; Musterd et al. 2017). To this aim, Southern and Northern European countries were considered as two regional contexts belonging to the same socioeconomic and demographic system (Rees et al. 2012). However, these two contexts diverge in demographic profiles, metropolitan hierarchies, socioeconomic structures, and labor market dynamics, so that they represent an interesting case of study (Potančoková et al. 2021).

Empirical evidence about long trends in the annual rate of change for both total and urban population tell us a sort of 'crispy tale' (Rodrigo-Comino et al. 2021). Urban population rate of change diverged in Northern and Southern Europe since the early 2000s (Salvati 2022). In Northern Europe, urban population growth substantially fueled total population growth, and the reverse holds for Southern Europe. This pattern can be associated with the better life conditions that normally characterize northern cities as greener, smarter, and with a higher level of well-being, than Southern cities (Pacione 2003; Salvati 2014; Rubiera-Morollon and Garrido-Yserte 2020). The growing level of spatial segregation of foreign population residing in Mediterranean cities, compared with the Northern European ones, and the growing level of spatial inequalities (Le Goix 2005; Gonzales 2011; Lee 2011) are examples of the emerging (and possibly, disadvantaged) conditions in these group of cities. These processes clearly act as detrimental to social cohesion and have negative effects on the sustainability of host societies (e.g. Gospodini 2009; Ferrara et al. 2016; Rontos et al. 2022).

By identifying individual processes of metropolitan expansion (e.g. Power 2001; Sassen 2010), a new vision for sustainable land management in local contexts under rapid socioeconomic modifications was derived and proposed (Salvati and Carlucci 2016). In particular, we consider the variation over time of the resident population

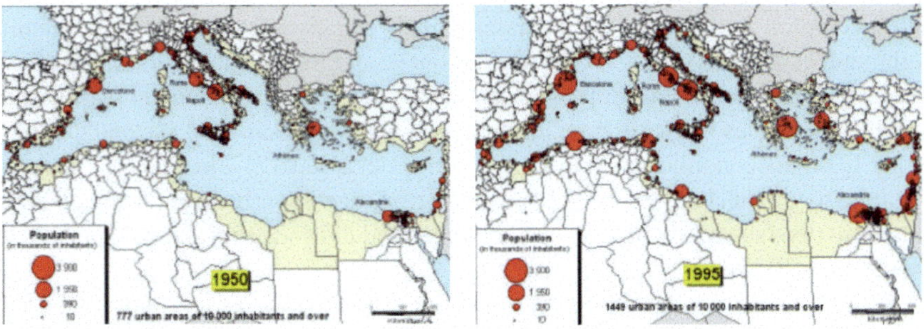

Fig. 3.10 Urban areas with more than 10 thousand inhabitants in Mediterranean coastal regions, comparing 1950 and 1995 status. *Source* Author's elaboration of United Nations Urbanization Prospects, demographic statistics, and cartographic databases

in Mediterranean cities (Fig. 3.10). Despite a relatively clear pattern stems from the maps, with relatively few changes over a sufficiently long time interval, further discussion is needed on the intrinsic diffusion of sparse, low-density settlements, and how—thanks to population deconcentration affecting both rural and urban areas—they have driven recent transformations in Mediterranean cities (e.g. Sauri 2003).

Nevertheless, these regions appear as suitable to understand long-term dynamics of discontinuous and dispersed urban expansion for three main reasons, as clearly recognized and discussed throughout this book. The first reason refers to urban growth policies depending on Mediterranean deregulation and informality, at least for a long-time window between the 1950s and the 1980s (Salvati et al. 2016). Second, less homogeneous demographic dynamics over time and space—as described by distinct waves of fast and slow population increase—have recently emerged and consolidated in the academic literature (Rontos et al. 2016). Third, a complex system of interacting agents with local peculiarities, independently from land prices and zoning processes, characterized 'Southern' urban expansion (Tomao et al. 2017).

3.6 Future Cities and the Legacy with Socioeconomic Crises and Environmental Challenges

The Lisbon strategy has put the need for sustainable growth at the heart of policy agenda in the old continent (Swyngedouw 2004). The European Union target of territorial cohesion was added to long-established objectives of economic and social cohesion (Samways 2022). In fact, the main demographic, economic, social, and environmental trends shaping European spatial development plans represent a challenge for a balanced and sustainable growth in the whole Union (Scott and Storper 2015). In this perspective, it was recognized widely how challenges such as regional socioeconomic disparities and pressures on the natural heritage (Fig. 3.11) require an integrated and multidisciplinary approach in order to correctly assess their impact, and a common policy response containing their negative externalities (Seifollahi-Aghmiuni et al. 2022). This is crucial in a context of climate change hitting distinctively urban and rural areas (Smiraglia et al. 2015). A rising demand for mitigation and adaptation policies to global transformations—spanning from land use and demography to socioeconomic structures (e.g. Ferrara et al. 2017)—justifies a common response to global change (Fig. 3.12), considering together urbanization and climate change as two faces of the same 'sustainability' medal (Lindstrom and Bartling 2003; Salvia et al. 2022).

Issues related to unbalanced resources, economic polarization, and territorial dichotomy along the Mediterranean basin are clearly the most significant to consider (e.g. Sallustio et al. 2018). Their assessment definitely requires an integrated, multidisciplinary approach (Quaas et al. 2007). In fact, socioeconomic vulnerability of a given municipality, district, or territory, is an issue that cannot be restricted to the resident population, while investing vast parts of the country itself (Onate and Peco 2005). The process often accelerates territorial unbalances, which may ultimately lead to social conflicts (Iosifides and Politidis 2005), enhancing migration movements (Pisani-Ferry 2005). This 'environmental spiral' may represent, in the coming future, a serious obstacle to the achievement of sustainable development in many dry areas of Mediterranean Europe (Salvati and Sateriano 2015).

Fig. 3.11 Soil impoverishment, water and wind erosion, and incipient land degradation under increasing climate aridity along Malaga fringe. *Source* Author's photographic archive

Fig. 3.12 A sealed section of Ilissos River in Kallithea, Central Athens, showing water pollution, landscape degradation, and sudden vulnerability to flooding. *Source* Author's photographic archive

We assume urbanization as one of the crucial issues of global change. Nowadays, we are increasingly observing a correlation between socioeconomic changes, urban land-use and landscape modifications (Salvati et al. 2018). Assuming urban expansion as gradually more dispersed and fragmented (Richardson and Chang-Hee 2004), topography and natural amenities were demonstrated to shape territorial development in suburban districts, in particular the expansion of urban settlements into high-quality landscapes (Polinesi et al. 2020). These are, in turn, associated with the preference for natural landscapes determining (or consolidating) income inequalities in local communities (Storper and Scott 2016). The resulting sociospatial polarization has been typically observed in Mediterranean cities under structural crises (Salvati et al. 2019). It results from intense recessions and continuous economic stagnation and may represent a concrete risk for sustainability and liveability of contemporary urban areas (Salvati et al. 2017). Therefore, it requires defined actions of spatial planning aimed at restoring the image of inclusivity and equality typical of the 'European city' style.

Official United Nations forecasts draw future scenarios with differential growth rates for urban and total population (Cabrè 2003). Even more marked will be the difference between the Southern European and Northern European contexts (Burillo et al. 2020). In the former case, total population will increase over time, but urban populations will also experience a slowdown in growth rates, which will nevertheless remain positive until the last period of observation (2050), when both urban and total population will have a negative growth rate (Boyle 2003). In the case of Northern Europe, the demographic growth pattern is predicted as becoming significantly different from the present time (Bocquier and Costa 2015). In fact, growth rates will be decreasing over time, while remaining substantially positive in relation to both total and urban population (Bocquier and Bree 2018).

Based on these premises, more flexible and generalized notions are required when evaluating urban expansion modes and designing well-informed socioterritorial policies that contrast unbalanced trajectories of regional development (Zambon et al 2019). In recent times, intensive land development with population decline became a common feature of a growing number of cities in advanced economies (Vinci et al. 2022). Therefore, providing evaluation instruments of the ongoing situation may support the decision-making process to improve the quality and equity of future urbanization in metropolitan areas around the world (Townsend 2009). Socially cohesive and spatially equitable cities seem to be the necessary antidote to multiple urban crises (e.g. economic, social, institutional, demographic, health) in Mediterranean regions (Tsilimingkas and Kizos 2014). Introducing a robust alternative to 'competitive models' of local development seems to be imperative in a context that reflects a sort of 'triple crunch' (namely, austerity policies, climate change, and increases in oil price). Going back to the local scale, abandoning interpretative paradigms oriented toward the logic of 'global networks' paralleled the idea of socioeconomic resilience (Ferrara et al. 2016; Tombolini et al. 2016; Wilson et al. 2017).

While resilience appears as a permanent issue in socioeconomic thought, institutions and policy makers have properly considered this notion in strategic planning and landscape design only in some cases (Zuindeau 2007). Dealing with resilience requires a deep analysis of societies, institutions, and local contexts, understanding the resilient nature beyond the system (Zwiers et al. 2016). This 'holistic' concept further complicates research on local factors of 'metropolitan resilience' and should be imperatively associated with the 'sustainability challenge' (Vinci et al. 2023). Under this viewpoint, however, resilience theory provides the appropriate knowledge and informs regional development built on local-based concepts, rather than focusing only on economic competitiveness factors (Tumpel-Gugerell and Mooslechner 2003). These policy perspectives are aimed at envisaging robust economic and social spaces, empowering producers and consumers, (i) to interact locally, (ii) to reduce dependency upon distant and larger scale agents, and, in turn, (iii) to directly shape the interaction between non-local (large) corporations and the nation state (Buzar et al. 2005).

3.7 Unraveling the Peculiarities of Contemporary Mediterranean Cities

Even if a number of Mediterranean cities seem to be competing to acquire a functional role in the global context, the stable entry of many of them into any system of 'global cities' still seems uncertain (Carbonaro et al. 2018). Uncertainties derive from the sedimentation of essential aspects of divergence with the more advanced metropolitan systems in the world (Carlucci et al. 2018), among which a significant role is played by the sociocultural mismatch between planning policies and urban practices (Fig. 3.13).

Within the region, the permanence of marked socioeconomic gaps between the metropolitan systems of the north bank and those of the south bank is also of great importance (Coleman 2006). Both problems are reflected in the inability to contain the attractive factors of large metropolitan areas on internal regions as well as their impact on surrounding regions (Ciommi et al. 2018). The demographic distance with the large European and American cities still appears as particularly marked (Oueslati et al. 2015). The same ranking of world cities ordered by demographic size may confirm this picture (Table 3.3). Only two Mediterranean cities (Cairo and

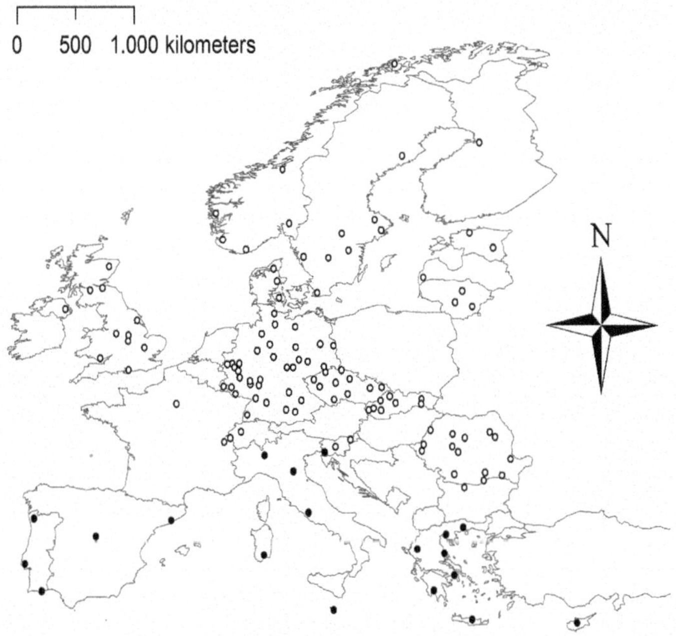

Fig. 3.13 Spatial distribution of large cities in Southern Europe (black dots) compared with the same class of cities (white dots) in the rest of Europe (Functional Urban Areas, FUA, sensu Eurostat classification). *Source* Author's elaboration on Urban Audit Eurostat shapefiles with countries' boundaries

Table 3.3 Demographic size (millions people) of the largest metropolitan agglomerations in the Mediterranean basin

City	Country	World rank	Inhabitants
El Cairo	Egypt	16	11.1
Istanbul	Turkey	22	9.7
Barcelona	Spain	51	4.8
Alexandria	Egypt	73	3.8
Rome	Italy	89	3.3
Athens	Greece	94	3.2
Algers	Algeria	95	3.2
Casablanca	Morocco	97	3.1
Tel Aviv	Israel	103	3.0
Aleppo	Syria	137	2.5
Smirne	Turkey	141	2.5
Damasco	Syria	153	2.3
Naples	Italy	159	2.2
Tunis	Tunisia	172	2.2
Tripoli	Libya	173	2.1
Beirut	Lebanon	207	1.8
Rabat	Morocco	225	1.6

Source Author's elaboration of United Nations demographic statistics and Urbanization Prospects, 2018

Istanbul) can boast a relevant global position in the world ranking, while the rest of the cities are positioned at least after the 50th position (this is the case of Barcelona). Cities like Rome and Athens are even lower in the ranking (Zambon et al. 2017).

A similar comparison, more specifically based on economic indicators, would not give better results. Among the most significant economic statistics, such as the Loughborough Group's special 'GaWC' classification aimed at identifying global cities—according to a rank from A (alpha world cities) to D (minimal evidence of world-city formation)—only nine Mediterranean urban agglomerations fall into this list; the first is Rome, placed in the seventh position (the second of the 'world cities rank') followed by Barcelona and Istanbul. Tel Aviv was placed in the following class (strong evidence of 'world-city formation'), while Athens, Lisbon, and Cairo (the only city on the Southern shore entering this class) were down in the ranking (some evidence in the 'formation of global cities').

The last class ('minimal evidence of world-city formation') includes two port cities, Genoa and Marseille, taken as centers of supra-regional importance experiencing partial decline (Hall and Pain 2006). Considering other economic indicators, the rank of Mediterranean cities in the global scene does not seem to change substantially (Kazepov 2005). The flows of economic transactions with London, considered as a honest expression of any global city, identified only three cities,

Barcelona, Istanbul, and Rome, with relatively low percentages compared with non-Mediterranean counterparts (Delladetsima 2012). On the impossibility of bringing the majority of Mediterranean urban contexts back towards the path drawn by global models, the words of Sassen (2001) seem to leave no margin for uncertainty:

> in the case of a complex landscape such as Europe's we see in fact several geographies of centrality, one global, other continental and regional (...). And then there are several geographies of marginality: the East-West divide and the North-South divide across Europe, as well as newer divisions. (...) Madrid, Barcelona, and Milan are gaining in the new European hierarchy; Naples, Rome, and Marseilles are not.

Globalization of social relations and labor markets, communication dematerialization, 'fluid' territories without clear borders and permeated with new forms of governance, have in turn established the importance of economic and cultural centers, meeting points, exchange of knowledge, flows intertwining, where 'similars' are recognized and 'strangers' are integrated within urban spaces (Cohen 2006). In this perspective, competition and the 'relational' sphere became the keywords of a process of progressive exit from marginality and semi-peripherality, to be part of an open and virtuous world system of social and economic relations (Delladetsima 2006). The shift from mono-centric (and often compact settlements) to widespread, branched, and polycentric cities has definitively marked, especially in the European experience, the rapid evolutionary sequence of modern cities (Fig. 3.14).

Such a perspective, however, cannot lead to a modification of the original idea of 'cities', transforming them from a mere 'planning object' to autonomous geographical units (Chorianopoulos et al. 2014). Determining new intersection points between global and local dynamics, even if inserted in 'secular' territorial contexts, may stimulate a rethinking of policies and planning at the various operational scales of observation, interpretation, and policy action (Beriatos and Gospodini 2004). In light of this transition, it is therefore appropriate to question, whether the destiny of Mediterranean cities will be marked by a progressive convergence towards urban forms, social styles, economic structures, and cultural conventions, similar to those observed in many cities in Northern Europe and the United States of America (Seltzer 2019). And if so, will this convergence be more or less spontaneous, more or less planned?

Keeping such motivations in mind, identification of basic elements framing Mediterranean cities, regions, and networks in a unitary perspective, is particularly appropriate at this point (Salvati 2016). Defining empirical models based on the comparison between metropolitan realities and regional specificities will contribute to the elaboration of a Mediterranean 'urban theory' that combines modern planning issues with concepts and experiences reflecting territorial specificities (Souliotis 2013). A quantitative analysis supporting such rationale should consider the characteristic land-use patterns of metropolitan regions (Catalàn et al. 2008; Schneider et al. 2010). In this perspective, an empirical exercise carried out on selected Mediterranean cities is provided in Fig. 3.15.

Results of the multidimensional ordination strategy based on multivariate statistics run on the percent share of selected land-use types in total landscape at the spatial level of Eurostat Functional Urban Areas (FUA), a spatial domain particularly appropriate in the assessment of socioeconomic dynamics and environmental

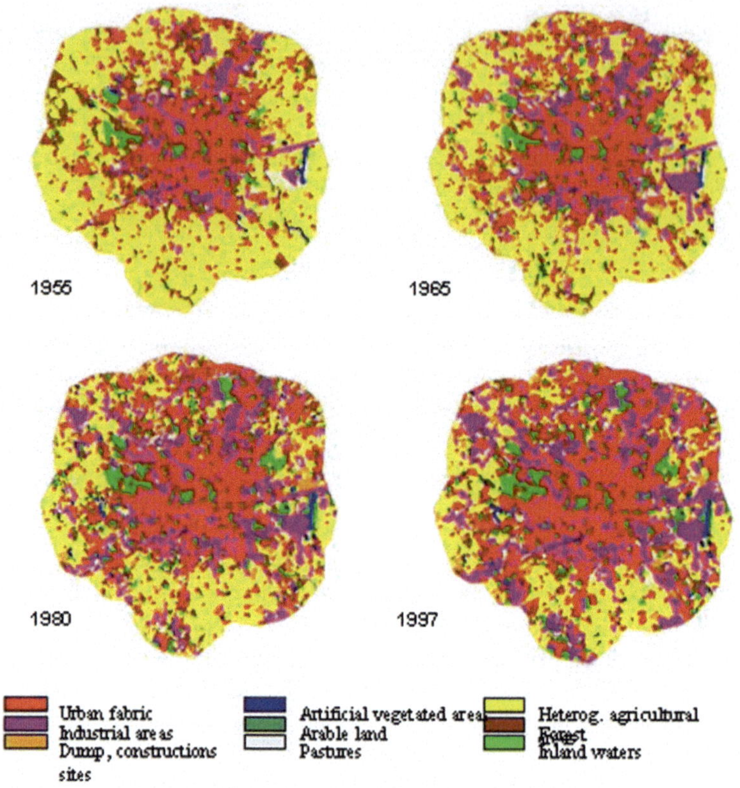

Fig. 3.14 Mono-centric expansion of urban settlements in Milan's municipality between 1955 and 1997, the time period reflecting the most intense compact growth in the area. *Source* Author's elaboration of EU-funded MOLAND project city profiles

state in metropolitan regions, are illustrated briefly in the following lines (Rodriguez-Rodriguez 2000). First, the ordination plot separates dense cities from urban agglomerations possibly displaying a more discontinuous settlement structure (Panori et al. 2018).

Second, compact settlements, dispersed settlements, and farmland were clearly separated along Components 1 and 2. Dense urban areas and cropland have displayed an opposite pattern, considering that Barcelona (Spain) and Athens (Greece) were the largest cities in Mediterranean Europe to be positively associated with the dimension of compact settlements, while Seville was a relatively large city in Southern Spain associated with the same dimension from the opposite side (cropland). Discontinuous settlements mixed preferentially with infrastructures, industrial centers, and green urban areas (Kourliouros 1997). Small cities in Italy, such as Verona, Acireale, and Ancona, were associated with this dimension. A rather large (and spatially dispersed) city such as Rome, was partly associated with this axis as well. These results finally reflect a well-defined profile of sprawl in originally compact cities (e.g. Galster

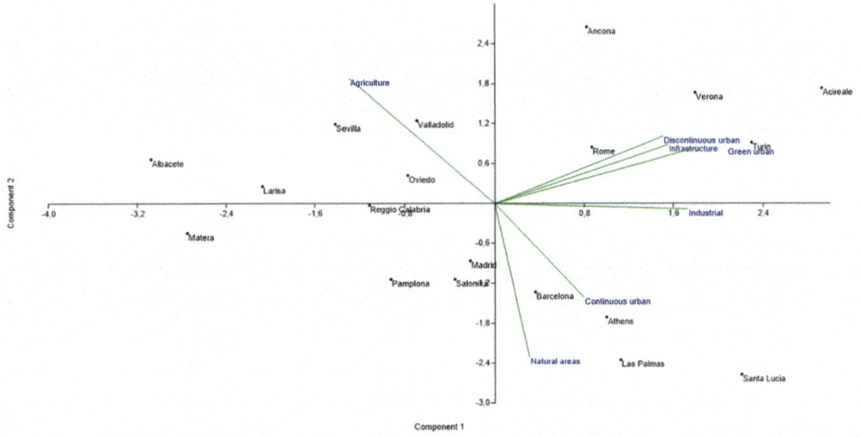

Fig. 3.15 Results of a multivariate ordination strategy; a biplot of Principal Component Analysis (loadings: land-use type vs. scores: cities) illustrating the specific landscape structure and characteristics of selected Mediterranean cities in Europe (explained variance: 41.6% on axis 1 and 30.5% on axis 2). *Source* Author's elaboration of Eurostat land-use data for individual cities derived from Copernicus Land Urban Atlas initiative of European Environment Agency

et al. 2001), and most characterized with specific forms of 'ribbon' expansion of mixed (residential and service) settlements in originally less compact and dense agglomerations (Giannakourou 2005).

3.8 Discussion

As outlined in Chapter 2 using a narrative approach, and more widely here in the present chapter adopting a quantitative perspective, urban growth has been extensively evaluated since World War II, with a wide range of scientific contributions focused on morphological and functional mechanisms of settlement expansion. Changes over time in urban forms, functional transformations involving local, urban, metropolitan, and regional scales, formation and consolidation of city networks and, more recently, the environmental implications of uncontrolled land consumption were debated widely in the academic literature. Until the 1970s, however, these contributions dealt primarily with the (ever-green) debate on compact (radio-centric) urban expansion following the vertical increase in dense or semi-dense settlements characteristic of both Western and Southern Europe. Jacobs (1961, p. 202) has provided a piece of literature when offering readers with a straightforward argumentation on the compact city, indicating that

> "city dwellings have to be intensive in their use of the land too, for reasons that go much deeper than cost of land. On the other hand, this does not mean that everyone can or should be put into elevator apartment houses to live – or into any other one or two types of dwellings. That kind of solution kills diversity by obstructing it from another direction. Dwelling densities

> are so important for most city districts, and for their future development, and are so little considered as factors in vitality (...). High dwelling densities have a bad name in orthodox planning and housing theory. They are supposed to lead to every kind of difficulty and failure, But in our cities, at least, this supposed correlation between high densities and trouble, or high densities and slums, is simply incorrect, as anyone who troubles to look at real cities can see."

Since the 1980s, a 'horizontal' settlement model has made its way on a global scale, involving areas where urban dispersion had took occasional ground in earlier decades, impacting social cohesion and economic dynamics of several cities. Under these interpretative lens, settlement diffusion has represented an academic issue of significant interest and with kaleidoscopic implications reflected in the tight interconnection with environmental issues and economic dynamics. As a matter of fact, factors leveraging metropolitan expansion since the 1980s have been increasingly and inherently related to multiple and diversified socioeconomic contexts, making the integrated analysis of settlement morphology and socioeconomic functions a complex challenge (Cohen 2006; Hall and Pain 2006; Angel et al. 2011).

To identify the main determinants of city growth, it is necessary to consider a number of fundamental factors and contextual indicators, including economic and demographic variables, the sociospatial structure, and territorial patterns, as well as institutional and cultural attributes (Kazepov 2005; Cassier and Kesteloot 2012; Bruegmann 2019). In Mediterranean cities, long-established metropolitan contexts remain heavily influenced by a complex interplay of these factors, as the region includes kaleidoscopic landscapes, urban forms, economic structures, and social contexts, which would justify the inherent difficulty of explaining and interpreting the uniqueness of each Mediterranean city within this complex panorama (Leontidou 1996). Based on these premises, a spatially unbalanced development is an obstacle to social cohesion, economic competitiveness, and demographic growth of any region (Kallis 2010). In this regard, the European Commission (1999, p. 21) argued that

> "a polycentric settlement structure across the whole territory of the EU with a graduated city-ranking must be the goal. This is an essential prerequisite for the balanced and sustainable development of local entities and regions and for developing the real location advantage of the EU vis-à-vis other large economic regions in the world."

In addition, those kinds of models are not compatible with the idea of a sustainable growth based upon the consolidation of a network of smart (and greener) cities of medium dimension well interconnected and suited for boosting processes of spatial (re)distribution of population (Kahn 2000). The estimates made available on both continental and national scales have delineated dispersed and geographically discontinuous urbanization, possibly associated with a 'spatially uncoordinated' settlement expansion, as a factor of economic decline and land degradation that is hard to revert and that should be faced with specific tactics grasping territorial peculiarities, and giving the right value to truly participated planning solutions.

Coastal urbanization and 'touristification', the progressive industrialization of inland districts, as well as the inexorable spread of infrastructures, all reflect the progressive shift of local economic systems—in Southern Europe as well as in other peripheral regions of the old continent—towards advanced services (Vidal et al. 2011). Changes in consumer preferences and lifestyles have given a significant input to territorial transformations from agricultural (low-density) settlements to modern systems of urban–rural places of economic production and consumption (Dijkstra et al. 2015). Referring to the original boundaries of compact cities, definitions of locations 'in' and 'out' are losing their geographical meaning (Vaiou 1997). The most recent metropolitan dynamics, and especially those moving in the direction of 'sprawl', suggest the importance urban (and rural) systems display a residual ability to preserve a basic 'resilience' level to intrinsic territorial transformations (Stergiouli and Hadjibiros 2012).

In this perspective, the chapter provides a framework interpreting urbanization dynamics in the Mediterranean region as possible examples applicable to other regions with similar socioeconomic characteristics (Salvati et al. 2013). An interpretation of Mediterranean urbanity is proposed according to local governance and socioeconomic structures, in a view to reduce regional disparities between urban and rural areas (e.g. Buckley 1998). This logical interpretation is grounded on the characteristics of the local context in view of the persistence of a deep crisis which affects the economy, institutions, and social classes. A discussion on future development trajectories of metropolitan regions should definitely stress short-term divergence processes, but also their long-term impact on Mediterranean spaces.

References

Adger WN, Crépin AS, Folke C, Ospina D, Chapin FS, Segerson K, Wilen J (2020) Urbanization, migration, and adaptation to climate change. One Earth 3(4):396–399

Alexandri G (2015) Reading between the lines: gentrification tendencies and issues of urban fear in the midst of Athens' crisis. Urban Stud 52(9):1631–1646

Allen J, Barlow J, Maloutas T, Padovan L (2004) Housing in southern Europe. Blackwell, London

Andreotti A, Le Galès P, Moreno Fuentes FJ (2013) Controlling the urban fabric: the complex game of distance and proximity in European upper-middle-class residential strategies. Int J Urban Reg Res 37(2):576–597

Angel S, Parent J, Civco DL, Blei A, Potere D (2011) The dimensions of global urban expansion: estimates and projections for all countries, 2000–2050. Prog Plan 75(2):53–107

Arapoglou VP, Sayas J (2009) New facets of urban segregation in Southern Europe. Eur Urban Reg Stud 16(4):345–362

Arbaci S (2008) (Re)viewing ethnic residential segregation in southern European cities: housing and urban regimes as mechanisms of marginalisation. Hous Stud 23(4):589–613

Arbaci S, Malheiros J (2010) De-segregation, peripheralisation and the social exclusion of immigrants: southern European cities in the 1990s. J Ethn Migr Stud 36(2):227–255

Balk DL, Deichmann U, Yetman G, Pozzi F, Hay SI, Nelson A (2006) Determining global population distribution: methods, applications and data. Adv Parasitol 62:119–156

Balsa-Barreiro J, Mendez M, Morales AJ (2022) Scale, context, and heterogeneity: the complexity of the social space. Sci Rep 12:9037

Beckers V, Poelmans L, Van Rompaey A, Dendoncker N (2020) The impact of urbanization on agricultural dynamics: a case study in Belgium. J Land Use Sci 15(5):626–643

Benassi F, Bonifazi C, Heins F, Lipizzi F, Strozza S (2020) Comparing residential segregation of migrant populations in selected European urban and metropolitan areas. Spatial Demogr 8:269–290

Benassi F, Naccarato A, Iglesias-Pascual R, Salvati L, Strozza S (2023) Measuring residential in multi-ethnic and unequal European cities. Int Migr 61(2):341–361

Benassi F, Salvati L (2020) Urban cycles and long-term population trends in a southern European City: a demographic outlook. Appl Spat Anal Policy 13(1):777–803

Beriatos E, Gospodini A (2004) "Glocalising" urban landscapes: Athens and the 2004 Olympics. Cities 21(3):187–202

Berisha E, Cotella G (2021) Towards sustainable urbanization. learning from what's out there. Land 10(4):356

Bianchini L, Egidi G, Alhuseen A, Sateriano A, Cividino S, Clemente M, Imbrenda V (2021a) Toward a dualistic growth? Population increase and land-use change in Rome. Italy. Land 10(7):749

Bianchini L, Marucci A, Sateriano A, Di Stefano V, Alemanno R, Colantoni A (2021b) Urbanization and long-term forest dynamics in a metropolitan region of southern Europe (1936–2018). Sustainability 13(21):12164

Biasi R, Brunori E, Smiraglia D, Salvati L (2015) Linking traditional tree-crop landscapes and agro-biodiversity in Central Italy using a database of typical and traditional products: a multiple risk assessment through a data mining analysis. Biodivers Conserv 24:3009–3031

Bocquier P, Bree S (2018) A regional perspective on the economic determinants of urban transition in 19th-century France. Demogr Res 38:1535–1576

Bocquier P, Costa R (2015) Which transition comes first? Urban and demographic transitions in Belgium and Sweden. Demogr Res 33:1297–1332

Boyle P (2003) Population geography: does geography matter in fertility research? Prog Hum Geogr 27(5):615–626

Bruegmann R (2019) Sprawl: a compact history. University of Chicago Press, Chicago

Buckley C (1998) Rural/urban differentials in demographic processes: the Central Asian states. Popul Res Policy Rev 17(1):71–89

Bueno-Suárez C, Coq-Huelva D (2020) Sustaining what is unsustainable: a review of urban sprawl and urban socio-environmental policies in North America and Western Europe. Sustainability 12(11):4445

Burillo P, Salvati L, Matthews SA, Benassi F (2020) Local-scale fertility variations in a low-fertility country: evidence from Spain (2002–2017). Can Stud Popul 47(4):279–295

Buzar S, Ogden P, Hall R, Haase A, Kabisch S, Steinfiihrer A (2007) Splintering urban populations: emergent landscapes of reurbanisation in four European cities. Urban Stud 44(4):651–677

Buzar S, Ogden PE, Hall R (2005) Households matter: the quiet demography of urban transformation. Prog Hum Geogr 29(4):413–436

Cabré A (2003) Facts and factors on low fertility in Southern Europe. The case of Spain. J Popul Social Sec S1:309–321

Cai Z, Liu Q, Cao S (2020) Real estate supports rapid development of China's urbanization. Land Use Policy 95:104582

Carbonaro C, Leanza M, McCann P, Medda F (2018) Demographic decline, population aging, and modern financial approaches to urban policy. Int Reg Sci Rev 41(2):210–232

Carlucci M, Chelli FM, Salvati L (2018) Toward a new cycle: short-term population dynamics, gentrification, and re-urbanization of Milan (Italy). Sustainability 10(9):3014

Cassiers T, Kesteloot C (2012) Socio-spatial inequalities and social cohesion in European cities. Urban Stud 49(9):1909–1924

Castells-Quintana D, Wenban-Smith H (2020) Population dynamics, urbanisation without growth, and the rise of megacities. J Dev Stud 56(9):1663–1682

Catalán B, Saurí D, Serra P (2008) Urban sprawl in the Mediterranean? Patterns of growth and change in the Barcelona Metropolitan Region 1993–2000. Landsc Urban Plan 85(3–4):174–184

Cazcarro I, Duarte R, Sánchez-Chóliz J (2012) Economic growth and the evolution of water consumption in Spain: a structural decomposition analysis. Ecol Econ 96:51–61
Chen W, Chi G (2022) Urbanization and ecosystem services: the multi-scale spatial spillover effects and spatial variations. Land Use Policy 114:105964
Chorianopoulos I, Pagonis T, Koukoulas S, Drymoniti S (2010) Planning, competitiveness and sprawl in the Mediterranean city: the case of Athens. Cities 27:249–259
Chorianopoulos I, Tsilimigkas G, Koukoulas S, Balatsos T (2014) The shift to competitiveness and a new phase of sprawl in the Mediterranean city: enterprises guiding growth in Messoghia-Athens. Cities 39:133–143
Ciommi M, Chelli FM, Carlucci M, Salvati L (2018) Urban growth and demographic dynamics in southern Europe: Toward a new statistical approach to regional science. Sustainability 10(8):2765
Cividino S, Halbac-Cotoara-Zamfir R, Salvati L (2020) Revisiting the "city life cycle": global urbanization and implications for regional development. Sustainability 12(3):1151
Cohen B (2006) Urbanization in developing countries: current trends, future projections, and key challenges for sustainability. Technol Soc 28(1–2):63–80
Colantoni A, Grigoriadis E, Sateriano A, Sarantakou E, Salvati L (2017) Back to Von Thunen: a southern European perspective on mono-centric urban growth, economic structure and non-urban land decline. Int Plan Stud 22(3):173–188
Coleman D (2006) Immigration and ethnic change in low-fertility countries: a third demographic transition. Popul Dev Rev 401–446
Couch C, Petschel-Held G, Leontidou L (2007) Urban sprawl in Europe: landscapes, land-use change and policy. Blackwell, Oxford
Cuadrado-Ciuraneta S, Durà-Guimerà A, Salvati L (2017) Not only tourism: unravelling suburbanization, second-home expansion and "rural" sprawl in Catalonia, Spain. Urban Geogr 38(1):66–89
Davis K (1955) The origin and growth of urbanization in the world. Am J Sociol 60(5):429–437
De Muro P, Monni S, Tridico P (2011) Knowledge-based economy and social exclusion: shadow and light in the Roman socio-economic model. Int J Urban Reg Res 35(6):1212–1238
Delladetsima PM (2006) The emerging property development pattern in Greece and its impact on spatial development. Eur Urban Reg Stud 13(3):245–278
Delladetsima PM (2012) Sustainable development and spatial planning: some considerations arising from the Greek case. Eur J Spat Dev 46:1–21
Di Feliciantonio C, Salvati L (2015) 'Southern' alternatives of urban diffusion: investigating settlement characteristics and socio-economic patterns in three Mediterranean regions. Tijdschr Econ Soc Geogr 106(4):453–470
Dijkstra L, Garcilazo E, McCann P (2015) The effects of the global financial crisis on European regions and cities. J Econ Geogr 15(5):935–949
Dombi M (2021) Types of planning systems and effects on construction material volumes: an explanatory analysis in Europe. Land Use Policy 109:105682
Domene E, Sauri D (2006) Urbanization and water consumption: influential factors in the metropolitan region of Barcelona. Urban Stud 43(9):1605–1623
Domene E, Sauri D, Parés M (2005) Urbanization and sustainable resource use: the case of garden watering in the metropolitan region of Barcelona. Urban Geogr 26(6):520–535
Durà-Guimera A (2003) Population deconcentration and social restructuring in Barcelona, a European Mediterranean city. Cities 20(6):387–394
Egidi G, Salvati L, Vinci S (2020) The long way to tipperary: city size and worldwide urban population trends, 1950–2030. Sustain Cities Soc 60:102148
Commission E (1999) European spatial development perspective: towards balanced and sustainable development of the territory of EU. Publications Office of EU, Luxembourg
Feng Y, Wang X, Du W, Liu J, Li Y (2019) Spatiotemporal characteristics and driving forces of urban sprawl in China during 2003–2017. J Clean Prod 241:118061

Ferrara A, Kelly C, Wilson GA, Nolè A, Mancino G, Bajocco S, Salvati L (2016) Shaping the role of 'fast' and 'slow' drivers of change in forest-shrubland socio-ecological systems. J Environ Manage 169:155–166
Ferrara C, Carlucci M, Grigoriadis E, Corona P, Salvati L (2017) A comprehensive insight into the geography of forest cover in Italy: exploring the importance of socioeconomic local contexts. Forest Policy Econ 75:12–22
Galster G, Hanson R, Ratcliffe MR, Wolman H, Coleman S, Freihage J (2001) Wrestling sprawl to the ground: defining and measuring an elusive concept. Hous Policy Debate 12(4):681–717
Gao M, Chen B, Xu Y, Li D (2023) Heterogeneous impacts of global land urbanization on land-use structure from economic and technological perspectives. Ecol Ind 147:109955
Garcia-López MÀ (2019) All roads lead to Rome... and to sprawl? Evidence from European cities. Reg Sci Urban Econ 79:103467
Garcia-López MA, Muñiz I (2010) Employment decentralisation: polycentricity or scatteration? The Case of Barcelona. Urban Stud 47(14):3035–3056
Giacalone M, Turco R, Mosconi EM, Alaimo LS, Salvati L (2023) The way toward growth: a time-series factor decomposition of socioeconomic impulses and urbanization trends in a pre-crisis European region. Social Ind Res 1–22
Giannakourou G (2005) Transforming spatial planning policy in Mediterranean countries. Europeanization and domestic change. Eur Plann Stud 13(2):253–264
Glaeser EL (2020) Urbanization and its discontents. East Econ J 46(2):191–218
Gonzales S (2011) The north/south divide in Italy and England: discursive construction of regional inequality. Eur Urban Reg Stud 18(1):62–76
Gospodini A (2009) Post-industrial trajectories of Mediterranean European cities: the case of post-Olympics Athens. Urban Stud 46(5–6):1157–1186
Grout C, Cavailhès J, Détang-Dessendre C, Thomas A (2016) Is sprawling residential behavior influenced by climate? Land Econ 92(2):203–219
Halbac-Cotoara-Zamfir R, Egidi G, Salvia R, Salvati L, Sateriano A, Gimenez-Morera A (2021) Recession, local fertility, and urban sustainability: results of a quasi-experiment in Greece, 1991–2018. Sustainability 13(3):1052
Halbac-Cotoara-Zamfir R, Marucci A, Salvia R, Quaranta G, Sateriano A, Cecchini M, Bianchini L (2022) Caring of the fringe? Mediterranean desertification between peri-urban ecology and socioeconomics. Sustainability 14(3):1426
Hall P, Pain K (2006) The polycentric metropolis. Learning from mega-city regions in Europe. Earthscan, London
Hamnett C (2020) Is Chinese urbanisation unique? Urban Stud 57(3):690–700
Henderson JV, Turner MA (2020) Urbanization in the developing world: too early or too slow? J Econ Perspect 34(3):150–173
Herrero-Jáuregui C, Concepción ED (2023) Effects of counter-urbanization on Mediterranean rural landscapes. Landscape Ecol 38(12):3695–3711
Iosifides T, Politidis T (2005) Socio-economic dynamics, local development and desertification in western Lesvos, Greece. Local Environ 10:487–499
Ismael HM (2021) Urban form study: the sprawling city—review of methods of studying urban sprawl. GeoJournal 86(4):1785–1796
Jacobs J (1961) The death and life of great American cities. Vintage Books, New York
James P (2014) Urban sustainability in theory and practice: circles of sustainability. Routledge, London
Jing C, Tao H, Jiang T, Wang Y, Zhai J, Cao L, Su B (2020) Population, urbanization and economic scenarios over the belt and road region under the shared socioeconomic pathways. J Geog Sci 30:68–84
Kahn M (2000) The environmental impact of suburbanization. J Policy Anal Manage 19(4):569–586
Kaika M (2003) Constructing scarcity and sensationalising water politics: 170 days that shook Athens. Antipode 35(5):919–954

Kallis G (2010) Coevolution in water resource development. The vicious cycle of water supply and demand in Athens, Greece. Ecol Econ 69(4):796–809

Karadimitriou N, Maloutas T, Arapoglou VP (2021) Multiple deprivation and urban development in Athens, Greece: spatial trends and the role of access to housing. Land 10(3):290

Kazepov Y (2005) Cities of Europe: changing contexts, local arrangements, and the challenge to urban cohesion. Blackwell, Oxford, UK

Keddie J, Tonkiss F (2010) The market and the plan: housing, urban renewal and socio-economic change in London. City Cult Soc 1(2):57–67

Kemeny T, Storper M (2012) The sources of urban development: wage, housing, and amenity gaps across American cities. J Reg Sci 52(1):85–108

Khan I, Hou F, Le HP, Ali SA (2021) Do natural resources, urbanization, and value-adding manufacturing affect environmental quality? Evidence from the top ten manufacturing countries. Resour Policy 72:102109

Kirk D (1996) Demographic transition theory. Popul Stud 50(3):361–387

Kockelkorn A, Schmid C, Streule M, Wong KP (2023) Peripheralization through mass housing urbanization in Hong Kong, Mexico City, and Paris. Plan Perspect 38(3):603–641

Koprowska K, Łaszkiewicz E, Kronenberg J (2020) Is urban sprawl linked to green space availability? Ecol Ind 108:105723

Kourliouros E (1997) Planning industrial location in Greater Athens: the interaction between deindustrialization and anti-industrialism during the 1980s. Eur Plan Stud 5(4):435–460

Kroll F, Kabisch N (2012) The relation of diverging urban growth processes and demographic change along an urban-rural gradient. Popul Space Place 18:260–276

Kwilinski A, Lyulyov O, Pimonenko T (2023) The effects of urbanisation on green growth within sustainable development goals. Land 12(2):511

Le Goix R (2005) Gated communities: sprawl and social segregation in southern California. Hous Stud 20(2):323–343

Lee S (2011) Metropolitan growth patterns and socio-economic disparity in six US metropolitan areas 1970–2000. Int J Urban Reg Res 35(5):988–1011

Leontidou L (1996) Alternatives to modernism in (Southern) urban theory: exploring in-between spaces. Int J Urban Reg Res 20(2):180–197

Stergiouli L, Martha HK (2012) The growing water imprint of Athens (Greece) throughout history. Reg Environ Change 12:337–345

Lindstrom MJ, Bartling H (2003) Suburban sprawl: culture, theory, and politics. Rowman & Littlefield, New York

Liu L, Meng L (2020) Patterns of urban sprawl from a global perspective. J Urban Plann Dev 146(2):04020004

Lowry IS (1990) World urbanization in perspective. Popul Dev Rev 16:148–176

Maloutas T (2004) Segregation and residential mobility. Socially entrapped social mobility and its impact on segregation in Athens. Eur Urban Reg Stud 11(2):171–187

Maloutas T (2007) Socio-economic classification models and contextual difference: the "European socio-economic classes" (ESeC) from a South European angle. South Eur Soc Polit 12(4):443–460

Marti-Hennberg J (2005) Empirical evidence of regional population concentration in Europe, 1870–200. Popul Space Place 11(4):269–281

Masini E, Tomao A, Barbati A, Corona P, Serra P, Salvati L (2019) Urban growth, land-use efficiency and local socioeconomic context: a comparative analysis of 417 metropolitan regions in Europe. Environ Manage 63:322–337

Menashe-Oren A, Bocquier P (2021) Urbanization is no longer driven by migration in low-and middle-income countries (1985–2015). Popul Dev Rev 47(3):639–663

Mouratidis K (2019) Compact city, urban sprawl, and subjective well-being. Cities 92:261–272

Mudu P (2006) Patterns of segregation in contemporary Rome. Urban Geogr 27(5):422–440

Mukhopadhyay P, Zérah MH, Denis E (2020) Subaltern urbanization: Indian insights for urban theory. Int J Urban Reg Res 44(4):582–598

Muñoz F (2003) Lock living: urban sprawl in Mediterranean cities. Cities 20(6):381–385
Murtin F (2013) Long-term determinants of the demographic transition, 1870–2000. Rev Econ Stat 95(2):617–631
Musterd S (2005) Social and ethnic segregation in Europe: levels, causes and effects. J Urban Aff 27(3):331–348
Musterd S, Marcińczak S, Van Ham M, Tammaru T (2017) Socioeconomic segregation in European capital cities. Increasing separation between poor and rich. Urban Geogr 38(7): 1062–1083
Nickayin SS, Tomao A, Quaranta G, Salvati L, Gimenez Morera A (2020) Going toward resilience? Town planning, peri-urban landscapes, and the expansion of Athens, Greece. Sustainability 12(24):10471
Oddo L, Zanini A (2022) The paradox of "Malthusian urbanization": urbanization without growth in the Republic of Genoa, 1300–1800. Eur Rev Econ Hist 26(4):508–534
Onate JJ, Peco B (2005) Policy impact on desertification: stakeholders' perceptions in southeast Spain. Land Use Policy 22:103–114
Orenstein DE, Frenkel A, Jahshan F (2014) Methodology matters: measuring urban spatial development using alternative methods. Environ Plann B Plann des 41:3–23
Oueslati W, Alvanides S, Garrod G (2015) Determinants of urban sprawl in European cities. Urban Studies 52(9):1594–1614
Pacione M (2003) Urban environmental quality and human wellbeing—a social geographical perspective. Landsc Urban Plan 65:19–30
Pandey B, Brelsford C, Seto KC (2022) Infrastructure inequality is a characteristic of urbanization. Proc Natl Acad Sci 119(15):e2119890119
Panori A, Psycharis Y, Ballas D (2018) Spatial segregation and migration in the city of Athens: investigating the evolution of urban socio-spatial immigrant structures. Popul Space Place e2209
Parés M, March H, Saurí D (2013) Atlantic gardens in Mediterranean climates: understanding the production of suburban natures in Barcelona. Int J Urban Reg Res 37:328–347
Pisani-Ferry J (2005) An agenda for a growing Europe. Oxford University Press, Oxford
Polinesi F, Recchioni M, Turco R, Rontos K, Rodrigo-Comino J, Benassi F (2020) Population trends and urbanization: simulating density effects using a local regression approach. ISPRS Int J Geo Inf 9(7):454
Potancokovà M, Stonawski M, Gailey N (2021) Migration and demographic disparities in macro-regions of the European Union, a view to 2060. Demogr Res 45:1317–1354
Power A (2001) Social exclusion and urban sprawl: is the rescue of cities possible? Reg Stud 35(8):731–742
Profiroiu CM, Bodislav DA, Burlacu S, Rădulescu CV (2020) Challenges of sustainable urban development in the context of population Growth. Eur J Sustain Dev 9(3):51–51
Quaas MF, Baumgartner S, Baker C, Frank K, Muller B (2007) Uncertainty and sustainability in the management of rangelands. Ecol Econ 62:251–266
Randolph GF, Storper M (2023) Is urbanisation in the global south fundamentally different? Comparative global urban analysis for the 21st century. Urban Stud 60(1):3–25
Rees P, Van der Gaag N, De Beer J, Heins F (2012) European regional populations: current trends, future pathways, and policy options. Eur J Popul 28(4):385–416
Richardson H, Chang-Hee C (2004) Urban sprawl in western Europe and the United States. Ashgate, London
Ritchie H, Roser M (2024) How urban is the world? Our world in data
Roberts S (1991) A critical evaluation of the city life cycle idea. Urban Geogr 12(5):431–449
Rodrigo-Comino J, Egidi G, Sateriano A, Poponi S, Mosconi EM, Gimenez Morera A (2021) Suburban fertility and metropolitan cycles: insights from European cities. Sustainability 13(4):2181
Rodriguez-Rodriguez V (2000) The seven demographic Spains. Inter-municipal contrasts in age structure: fertility and migration are the determining factors. Espace Popul Soc 18(3):425–435

Rontos K, Grigoriadis S, Sateriano A, Syrmali M, Vavouras I, Salvati L (2016) Lost in protest, found in segregation: divided cities in the light of the 2015 'Oki' Referendum in Greece. City Cult Soc 7(3):139–148
Rontos K, Papazoglou M, Syrmali ME, Sateriano A (2022) Towards the eastern gate to Europe: factors shaping perceptions and attitudes towards migrants in Lesvos Island (Greece), 2016. Reg Stat 12(1):104–134
Rubiera-Morollón F, Garrido-Yserte R (2020) Recent literature about urban sprawl: a renewed relevance of the phenomenon from the perspective of environmental sustainability. Sustainability 12(16):6551
Sahana M, Ravetz J, Patel PP, Dadashpoor H, Follmann A (2023) Where is the peri-urban? A systematic review of peri-urban research and approaches for its identification and demarcation worldwide. Remote Sens 15(5):1316
Sallustio L, Pettenella D, Merlini P, Romano R, Salvati L, Marchetti M, Corona P (2018) Assessing the economic marginality of agricultural lands in Italy to support land use planning. Land Use Policy 76:526–534
Salvati L (2014) Towards a polycentric region? The socioeconomic Trajectory of Rome, an 'eternal' Mediterranean city. Tijdschr Econ Soc Geogr 105(3):268–284
Salvati L, Carlucci M (2016) Patterns of sprawl: the socioeconomic and territorial profile of dispersed urban areas in Italy. Reg Stud 50(8):1346–1359
Salvati L, Ciommi MT, Serra P, Chelli FM (2019) Exploring the spatial structure of housing prices under economic expansion and stagnation: the role of socio-demographic factors in metropolitan Rome, Italy. Land Use Policy 81:143–152
Salvati L, Ferrara A, Chelli F (2018) Long-term growth and metropolitan spatial structures: an analysis of factors influencing urban patch size under different economic cycles. Geografisk Tidsskrift—Dan J Geogr 118(1):56–71
Salvati L, Morelli VG, Rontos K, Sabbi A (2013) Latent exurban development: city expansion along the rural-to-urban gradient in growing and declining regions of southern Europe. Urban Geogr 34(3):376–394
Salvati L, Sateriano A (2015) In-between coast and inland: a long-term assessment of demographic dynamics, agriculture and urban growth in Rome prefecture. Rendiconti Lincei 26:605–614
Salvati L (2022) Endogenous population dynamics and metropolitan cycles: long-term evidence from Athens, and eternally Mediterranean city. Eur J Popul 38(5):835–886
Salvati L (2016) The dark side of the crisis: disparities in per-capita income (2000–2012) and the urban-rural gradient in Greece. Tijdschr Econ Soc Geogr 107(5):628–641
Salvati L, Sateriano A, Grigoriadis E, Carlucci M (2017) New wine in old bottles: the (changing) socioeconomic attributes of sprawl during building boom and stagnation. Ecol Econ 131:361–372
Salvati L, Venanzoni G, Serra P, Carlucci M (2016) Scattered or polycentric? Untangling urban growth in three southern European metropolitan regions through exploratory spatial data analysis. Ann Reg Sci 57:1–29
Salvia R, Quaranta V, Sateriano A, Quaranta G (2022) Land resource depletion, regional disparities, and the claim for a renewed 'sustainability thinking'under early desertification conditions. Resources 11(3):28
Samways D (2022) Population and Sustainability: reviewing the relationship between population growth and environmental change. J Popul Sustain 6(1):15–41
Sassen S (2010) The city: its return as a lens for social theory. City Cult Soc 1(1):3–11
Saurí D (2003) Lights and shadows of urban water demand management. The case of the metropolitan region of Barcelona. Eur Plann Stud 11(3):229–243
Schneider F, Kallis G, Martinez-Alier J (2010) Crisis or opportunity? Economic degrowth for social equity and ecological sustainability. Introduction to this special issue. J Clean Prod 18(6):511–518
Scott AJ, Storper M (2015) The nature of cities: the scope and limits of urban theory. Int J Urban Reg Res 39(1):1–15

Seifollahi-Aghmiuni S, Kalantari Z, Egidi G, Gaburova L, Salvati L (2022) Urbanisation-driven land degradation and socioeconomic challenges in peri-urban areas: Insights from Southern Europe. Ambio 51(6):1446–1458

Seltzer N (2019) Beyond the Great Recession: labor market polarization and ongoing fertility decline in the United States. Demography 56(4):1463–1493

Shaw BJ, van Vliet J, Verburg PH (2020) The peri-urbanization of Europe: a systematic review of a multifaceted process. Landsc Urban Plan 196:103733

Smiraglia D, Ceccarelli T, Bajocco S, Perini L, Salvati L (2015) Unraveling landscape complexity: land use/land cover changes and landscape pattern dynamics (1954–2008) in contrasting peri-urban and agro-forest regions of northern Italy. Environ Manage 56:916–932

Solly A, Berisha E, Cotella G, Janin Rivolin U (2020) How sustainable are land use tools? A Europe-wide typological investigation. Sustainability 12(3):1257

Souliotis N (2013) Cultural economy, sovereign debt crisis and the importance of local contexts: the case of Athens. Cities 33:61–68

Storper M, Scott AJ (2016) Current debate in urban theory: a critical assessment. Urban Stud 53(6):1114–1136

Sun L, Chen J, Li Q, Huang D (2020) Dramatic uneven urbanization of large cities throughout the world in recent decades. Nat Commun 11(1):5366

Swyngedouw E (2004) Social power and the urbanization of water. Oxford University Press, Oxford

Tian Y, Tsendbazar NE, van Leeuwen E, Fensholt R, Herold M (2022) A global analysis of multi-faceted urbanization patterns using Earth Observation data from 1975 to 2015. Landsc Urban Plan 219:104316

Tomao A, Quatrini V, Corona P, Ferrara A, Lafortezza R, Salvati L (2017) Resilient landscapes in Mediterranean urban areas: understanding factors influencing forest trends. Environ Res 156:1–9

Tombolini I, Colantoni A, Renzi G, Sateriano A, Sabbi A, Morrow N, Salvati L (2016) Lost in convergence, found in vulnerability: a spatially-dynamic model for desertification risk assessment in Mediterranean agro-forest districts. Sci Total Environ 569:973–981

Townsend A (2009) Integration of economic and spatial planning across scales. Int J Public Sect Manag 22(7):643–659

Tsilimigkas G, Kizos T (2014) Space, pressures and the management of the Greek landscape. Geografiska Annaler Ser B Human Geogr 96(2):159–175

Tumpel-Gugerell G, Mooslechner P (2003) Economic convergence and divergence in Europe. Growth and regional development in an enlarged European Union. Edward Elgar, London

Vaiou D (1997) Facets of spatial development and planning in Greece. Eur Plan Stud 5(4):431–433

Vidal M, Domene E, Saurí D (2011) Changing geographies of water-related consumption: residential swimming pools in suburban Barcelona. Area 43(1):67–75

Vinci S, Egidi G, López Gay A, Salvati L (2022) Population growth and urban management in metropolitan regions: the contribution of natural balance and migration to polycentric development in Barcelona. Appl Spat Anal Policy 15(1):71–94

Vinci S, Vardopoulos I, Salvati L (2023) A tale of a shrinking City? Exploring the complex interplay of socio-demographic dynamics in the recent development of Attica, Greece. Cities 132:104089

Wang Q, Lin J, Zhou K, Fan J, Kwan MP (2020) Does urbanization lead to less residential energy consumption? A comparative study of 136 countries. Energy 202:117765

Wang Q, Wang X, Li R (2022) Does urbanization redefine the environmental Kuznets curve? An empirical analysis of 134 Countries. Sustain Cities Soc 76:103382

Wilson GA, Kelly CL, Briassoulis H, Ferrara A, Quaranta G, Salvia R, Detsis V, Curfs M, Cerda A, El-Aich A, Liu H, Kosmas C, Alados CL, Imeson A, Landgrebe-Trinkunaite R, Salvati L, Naumann S, Danwen H, Iosifides T, Kizos T, Mancino G, Nolè A, Jiang M, Zhang P (2017) Social memory and the resilience of communities affected by land degradation. Land Degrad Dev 28(2):383–400

Xiong W, Chen B, Wang H, Zhu D (2020) Public–private partnerships as a governance response to sustainable urbanization: lessons from China. Habitat Int 95:102095

Zambon I, Colantoni A, Salvati L (2019) Horizontal vs vertical growth: understanding latent patterns of urban expansion in large metropolitan regions. Sci Total Environ 654:778–785

Zambon I, Serra P, Sauri D, Carlucci M, Salvati L (2017) Beyond the 'Mediterranean city': socioeconomic disparities and urban sprawl in three Southern European cities. Geografiska Annaler Ser B Human Geogr 99(3):319–337

Zhang L, Yang L, Zohner CM, Crowther TW, Li M, Shen F, Zhou C (2022) Direct and indirect impacts of urbanization on vegetation growth across the world's cities. Sci Adv 8(27):eabo0095

Zhang W (2008) A forecast analysis on world population and urbanization process. Environ Dev Sustain 10:717–730

Zuindeau B (2007) Territorial equity and sustainable development. Environ Values 16:253–268

Zwiers M, Bolt G, van Ham M, Van Kempen R (2016) The global financial crisis and neighbourhood decline. Urban Geogr 37(5):664–684

Chapter 4
Metropolitan Sustainability, Urban Cultures, and the Basis for Mediterranean 'Sprawled Cities'

Abstract Urbanization is a powerful expression of the cultural traits of any human society. However, a comprehensive picture of Mediterranean urbanities has been rather difficult to delineate, as it has experienced recurrent conflicts and intense exchanges from three different continents. This contribution reviews and debates the dominant 'urban cultures' in post-war Southern European cities. The empirical findings of our study—corroborated with an extensive literature review from multiple disciplinary perspectives—reveal that the Mediterranean region results as heterogeneous in some aspects (e.g. spatial forms), and more similar in others (e.g. social dynamics and behaviors). Rejecting a unique model at the base of the 'Mediterranean city' archetype does not imply that there are no common characteristics that metropolitan centers of this geographical region may share. We finally focused on the peculiar dynamics and attitudes of the Mediterranean societies which have played, together with informal economic systems, a crucial role in the metropolitan development of the region.

Keywords Social diversification · Economic informality · Housing · Planning · Morphology · Mediterranean

4.1 Introduction

Sustainable development is a transversal concept to many academic disciplines in both socioeconomic and environmental fields. By interconnecting ecological and demographic issues—as well as the dimensions of economics, culture, technology, and institutions—sustainable development has been considered, in recent decades, the organizing principle for the achievement of any human development goal, simultaneously supporting the capacity of local systems to provide natural resources and ecosystem services, on which the economy may depend (Glaeser 2020; Hamnett 2020; Profiroiu et al. 2020). The first result of a long-term path oriented toward sustainable development is a society where living conditions and resources are used

L. Salvati, *Economic Growth, Regional Disparities, and Urban Sprawl*, Springer Geography, https://doi.org/10.1007/978-3-031-85799-7_4

to continue satisfying human needs without compromising the integrity and stability of natural systems (Shaw et al. 2020).

The motivation behind the (growing) scientific and political interest in sustainability issues comes from the awareness of the important technological, production, organizational, and labor law challenges to which policy makers, associations representing jobs and businesses, environmental organizations and civil society, are called to respond with the objective of achieving a greater competitiveness in national productions (Khan et al. 2021). The transformations of international markets and, together, pressing obligations and recommendations on sustainability at all governance scales—and the dimensions of intervention that arise from decisions at both European and global levels—are leveraging such challenges (Sun et al. 2020a). To this end, mainstream research envisages the 'sustainability' paradigm in a mostly operational way, requiring systematized information that includes regulatory frameworks, a deep analysis of social, territorial, and corporate bargaining platforms, as well as contextual quali-quantitative data and indicators (e.g. Cai et al. 2020), based on an extensive review of academic and political debates.

With this perspective in mind, a comprehensive literature review usually proceeds through a broad, and reasoned, scrutiny of the secondary normative-regulatory sources on the subject of environmental, social, and economic sustainability at global, European, and national levels (Xiong et al. 2020). A trans-scalar vision oriented toward theoretical and operational studies identifying the most significant agents at the various policy levels seems to be particularly appropriate (Sun et al. 2020b). At the same time, a historical analysis of the concept of 'sustainable development' on a global scale (sensu Wang et al. 2020), starts from the experience of the 1960s environmental movements in the United States, addressing the 'limits to development' typical of the 1970s positivistic debate, and introducing perhaps the most relevant document on such issue, namely the Bruntland Report dated 1987 (Adger et al. 2020; Oddo and Zanini 2022; Tian et al. 2022).

The following time period reflected a progressive refinement in the concept of 'sustainability', which was represented no longer as a process of social development, economic growth, and normative protection of the environment, but as a real style of collective and individual life (Zhang et al. 2022). This has led to the drafting of Agenda 21, the millennium development goals and, more recently, Agenda 2030, with a dominant social target, and a concrete view to containing climate change (Pandey et al. 2022). From this point of view, integrated social, economic, and ecological issues become crucial to any sustainable development strategy and, as such, they have been tackled in their various meanings, as a unitary and interconnected dimension with the other objectives of the 2030 Agenda on a global scale (Castells-Quintana and Wenban-Smith 2020).

While exploring carefully the regional context, literature focusing on the European level also provides an overview of the integrated strategies of sustainable development (Murtin 2013), reflecting one of the main community objectives that is consolidating over time, namely the transition toward decarbonization and a climate-neutral economy (Gao et al. 2023). At the national level, scholars, experts, and practitioners, should increasingly rethink the recent stages of sustainable development

Fig. 4.1 Sustainable Development Goals (SDGs); SDG11 was specifically addressing 'sustainable cities and communities' (dashed circle). *Source* Author's elaboration of Agenda 2030 scheme

(e.g. Cuadrado-Ciuraneta et al. 2017). Having full adherence to the objectives of the 2030 Agenda (Fig. 4.1) and the European energy strategy (Sumari et al. 2020), the most recent developments—understood as a priority for all the Member Countries—should be further discussed from the perspective of the community 'new green deal'.

Ultimately, the notion of sustainable development and, more generally, the 'holistic' paradigm of sustainability, has represented increasingly dominant concepts in the debate on economic growth and the competitiveness of production systems, becoming crucial elements in every strategy addressing just and equitable societies (Angel et al. 2011; Koprowska et al. 2020; Liu and Meng 2020). A rising awareness of the environmental dimension has gradually accompanied the historical evolution of (and the academic debate on) such issues (Bianchini et al. 2021). Environmental quality has considered a necessary dimension in the debate on long-term development paths that are both durable and ecologically 'neutral' (Cazcarro et al. 2012; Ismael 2021; Kundu et al. 2023). This reflects a developmental process that may bring benefits and minimize negative externalities to the ecosystems on which (and often thanks to which) societies live and thrive (Townsend 2009; Vidal et al. 2011; Andreotti et al. 2013). The objective of eliminating (or reducing) the negative externalities of socioeconomic systems on the environment has therefore been added to the long-lasting debate on growth.

4.2 The (Semantic) Linkage Between Cities and Sustainability

In a true vision of 'sustainable development', 'cities' should be immediately connected with the notion of 'civilization' (Bruegmann 2019). Namely, the former delineates the structure of human societies in terms of, e.g. spatial patterns of urbanization/suburbanization, housing typologies and preferences, regional dynamics of population growth and decline), while the latter specifically refers to the humans that are part of the society, and their intangible characteristics, such as culture, beliefs, religion attitudes, and so on (Arapoglou and Sayas 2009). Therefore, speaking about 'cities' may evoke, together, the notion of a 'social life', opposed to a rural, more isolated life, and the modalities with which humans inhabit the territory, and mainly urban spaces (Fig. 4.2). In this perspective, urbanization processes (more or less directly) reflect cultural and intellectual characteristics of a given society (Arbaci and Malheiros 2010).

Metropolitan growth thus evolves in response to various factors of change (Menashe-Oren and Bocquier 2021) and basically as a product of the joint action of kaleidoscopic social players (Arbaci 2008), whose impact on landscapes and territories may significantly vary over time and space (Cassiers and Kesteloot 2012). Unique development types characteristic of a (supposedly) uniform European space, is clearly hard to be identified and qualified, as several basic elements—from vastly different disciplinary dimensions—may vary significantly between the countries (and the regions) of the same continent (Buzar et al. 2007; Chorianopoulos et al. 2010, 2014; Bueno-Suarez and Coq-Huelva 2020).

In the past, little attention had been given in trying to systematically explain urban structures through the relationship of social classes within the city, e.g. in light of metropolitan sustainability (Fig. 4.3). In spontaneous alternative cultures, 'where informality, community life and socializing, song and football attendance, mutual aid and illegal building are the main characteristics' (Leontidou 1990), a different approach from the traditional one can be adopted in order to analyze land-use patterns. Under this assumption, Leontidou (1996) proposed a research method investigating social dynamics of Mediterranean populations (Di Feliciantonio and Salvati 2015). The present work will take some advantages from Leontidou's human agency investigation in order to describe how Mediterranean urbanization patterns evolve (Durà-Guimera 2003).

This piece of work introduces a brief analysis of the main factors at the base of post-war metropolitan development in the old continent, and more specifically in Southern Europe. We converge on the opportunity of taking the Mediterranean basin as an independent area from the rest of Europe, which may require a dedicated approach (Newman and Thornely 1996), since the history of civilization in this region was extremely long and articulated (Fig. 4.4). Generally speaking, most of the existing Mediterranean cities were, and still are, places where history, culture, and local traditions are intertwined into a unique 'territorial context', that was (and

Fig. 4.2 Moving from Avinguda Meridiana toward the city center, a popular manifestation invades downtown Barcelona. *Source* Author's elaboration from press agencies

Fig. 4.3 From sprawl to desertification: examples of land degradation in fringe Athens (left), and Rome (right). *Source* Author's photographic archive

Fig. 4.4 Kaleidoscopic morphology of suburban settlements in Rome, five kilometers south of the historical center. *Source* Author's elaboration on Google Earth imagery

still is largely) influenced by the geographical and climatic characteristics stemming from the strong connection with the sea basin.

No doubt that several cityscapes in the area have been often defined with architectonical styles that reflect the heritage of different civilizations—consolidating multiple and often kaleidoscopic designs based on continuous stratifications and addictions of settlement layers. Being a crossroad of cultures, where the coexistence of different influences has given rise to an environment rich in history and diversity, Southern European cities were frequently oriented toward the sea, at least in an indirect way. With their narrow streets, shady alleys, lively squares—hopefully with more green in the future—and houses with big terraces overlooking the sun, Mediterranean cities—and especially the coastal ones—create an atmosphere that mixes the 'serenity of nature' with the energy of urban life, in line with what Albert Camus, the French man of letters lent to the Mediterranean literature, tried to express in most of his masterpiece production.

Based on this reflection, the main spatial patterns characteristic of Mediterranean cities in the last century were identified, delineated and briefly analyzed, including (i) what is frequently called with the 'Inverse-Burgess' model (Leontidou 1996), (ii) the peculiarity of a 'niche' phenomenon such as 'vertical differentiation'—better known with the term of 'vertical segregation' (Gospodini 2009), and (iii) a sort of 'popular land control', e.g. manifested through spontaneous building activity and the consolidation of 'informal neighborhoods', basically out-of-plans, or just tolerated by a permissive regime of spatial planning (Costa et al. 1991), more interested to accommodate votes and consensus to local parties than to preserve territorial traditions and cultural landscapes (Fig. 4.5).

Fig. 4.5 Popular creativity, the supremacy of social networks, and the multidimensional notion of 'Mediterranean koinonia': the city life of white collars striking in downtown Athens. *Source* Author's photographic archive

The following section proposes a specific analysis of Mediterranean urban development paths as a result of peculiar social-demographic processes, production modes, and cultural models (Mudu 2006; De Muro et al. 2011; Karadimitriou et al. 2021). A special effort was deserved to delineate the strategic role of subordinate economic forces (informal economy), their main player (low social classes shaping the dominant modes of production), and the spatial pattern of class and racial segregation. The last part of this contribution introduces and discusses on a presumably broad notion concerning a unique 'urban ideology' characteristic of a (supposedly homogeneous) Mediterranean region (Handlin 1963; King et al. 1997), as an original contribution to understanding and planning (future) Mediterranean cities (Kaika 2003; Masini et al. 2019; Gross and Ouyang 2021).

4.3 Mediterranean Urbanities and Metropolitan Cultures

A multitude of conceptually different criteria was adopted to define, delimitate, and qualify the Mediterranean region (Krumholz 1992; Munoz 2003; Musterd 2005), and most of them are grounded on historical similarities between countries, climatic conditions, geographical characteristics, current socioeconomic settings, and more generally, a sort of 'territorial amalgam' (Balchin 1996). To identify this region, it is important to consider both natural and cultural boundaries (e.g. Salvati et al. 2013). The area is clearly defined by a climate characterized by hot, dry summers and mild, rainy winters, which contribute to a unique ecosystem. Culturally, the region is a crossroad of ancient traditions, reflecting the influence of several historical civilizations, which help to trace the boundaries of one of the richest areas of cultural diversity and heritage in the world. In this perspective, homogeneous cultivation of crops as a source of food and fibers, consumption habits, mental and cultural tools, language, religion, and laws were frequently cited as basic working criteria (Braudel 1987), and one of the most famous definition of the Mediterranean as the 'old world region where olive trees grow' (Braudel 1953), seems to remain written on stones. However, this will not prevent that, throughout history, a long series of conflicts for the domain over the region have been frequent and intense (Fig. 4.6).

Opposing to a 'romantic' view of a 'unique Mediterranean mood' (Salvati and Carlucci 2016), the region resulted to be heterogeneous as a result of the multitude of 'collisions' between civilizations (Sassen 2010), making the identification of a single model of urbanization for the whole area a particularly hard and challenging task (Ferras 1977; Seronde Babonaux 1983; Rontos et al. 2016). Assuming Mediterranean cities as undergoing diverging urbanization patterns along their millenary history (e.g.

Fig. 4.6 Dusk at Salamina strait in a shot from Perama, Western Athens. *Source* Author's photographic archive

Parés et al. 2013; Salvati 2014; Tsilimigkas and Kizos 2014), four dominant urban typologies can be distinguished, namely (i) the Roman city, (ii) the Hellenistic city, (iii) the Byzantine city, and (iv) the Arab city (Spina 2004).

While some scholars believe that the limitations for studying the Mediterranean region as a whole mainly rely in the diverging characteristics of its cities, metropolitan areas, and the surrounding landscapes (e.g. Garcia-Lopez 2019), others seem to be more attracted and, possibly, fascinated by such kind of diversities (e.g. Leontidou 1990). In this perspective, the original traits of the Mediterranean region can only be explained having in mind the interaction of physical and anthropological drivers of change that have shaped societies since millennia (Domene et al. 2005), and that require the dismission of a 'linear' logic when interpreting its intimate nature (Fig. 4.7).

'Linear' and strictly 'quantitative' approaches—mainly based on statistics and indicators, primarily depicting the evolution of the economic domain—will inevitably lead to a 'reduced' and 'marginalized' understanding of Mediterranean spaces (Fried 1973). For this reason, in order to delineate metropolitan dynamics, 'qualitative' approaches should be adopted considering human-spatial relations (Maloutas 2007). According to Camus, the Mediterranean tradition is characterized by a 'solar ideal' manifested in the society by a reflexive liberty, an individualistic altruism that privileges elements such as limits and nature over excess and

Fig. 4.7 Panoramic view of downtown Granada dominated by Ahlambra Muslim fortification, from San Pedro hill. *Source* Author's photographic archive

history (Giaccaria and Minca 2011). Families stand out as the principal element of Mediterranean societies from Gramsci's analysis on Southern Italy (De Rosa and Salvati 2016), indirectly explaining the feeble urbanization of the Mediterranean, where communities' well-being is possibly weak, thanks to congestion esternalities (Maloutas 2004).

The strong relationship between relatives has influenced the distribution of social groups on the territory shaping, at least indirectly, the spatial organization of Mediterranean cities (Fig. 4.8). Cassano finally highlighted how the respect of rules, typical of central European populations, is contradicted in the Mediterranean by the strong attitude toward 'dispensation' (Bialasiewicz et al. 2013). This assumption reflects a 'historical' interpretation of the Mediterranean basin as a place where 'the alliance between man and nature derives from the Greek myths and divinities' (Bialasiewicz and Minca 2005).

Fig. 4.8 Urban growth, local communities, and harsh landscapes: settlement neighborhoods of Tourkovounia badlands, downtown Athens. *Source* Author's photographic archive

4.4 'Urban' Versus 'Metropolitan': Rethinking Mediterranean Geographies

Social processes constructed a strong identity of Mediterranean urban spaces—basically through architectural solutions quite distant from those promoted in a 'northern' mainstream. Following Leontidou (1996), this condition was barely interpreted as a sort of 'alternative modernity' responding to the rapid urbanization of the 1960s and the 1970s. Acting on heterogeneous territories with fragmented land ownership, urbanization allowed petty bourgeoisie consolidating and claiming for a non-subordinate role in the socioeconomic panorama of Mediterranean cities. These dynamics—highly stratified and possibly sedimented over time—are reflected in compact and dense morphologies, whose identitarian traits all over the region have only recently been questioned, following a pervasive settlement dispersion emerging since the 1990s (Hall 1997).

In this context, sprawl has led to a sort of a 'new modernity' understood as a cultural alternative to traditional modernism—in-between the outcome of (Western) European (e.g. French) centralism and the 'one-way' visions of American planning (Kroll and Kabisch 2012). Urban spaces have been incubators of this culture for years, shaping it through local experiences, landscapes, policy solutions, sometimes backward, often creative (Lee 2011). At the same time, it is difficult to find a comparative example of the pervasive social mix in land tenure and housing ownership—as well as in the use of public space—typical of Mediterranean cities, when investigating Western and/or Northern counterparts (Grout et al. 2016).

Self-promotion, informality, spontaneism, lasses-faire planning, deregulation, are different sides of the same medal (Le Goix 2005) that represents a privileged object of the dialogue between local development and the Mediterranean scale. Reflecting a continuous oscillation between emergency and consolidation (Kemeny and Storper 2012), this dialogue intimately represents the continuous tension between economic marginality and creativity, and between modernism and traditionalism, two elements of Mediterranean urban cultures and traditions (Zapata-Barrero 2023). In this perspective, overpopulated agglomerations of the past determined the shapes of today's modern cities (Power 2001).

'Overurbanization' and real estate speculation were general trends in the region, not only in the far past, but also in the last century, delineating a dominant and common pattern for several decades, at least between the mid-1940s and the late 1980s (Musterd et al. 2017). Most of the responsibility for such expansion patterns can be attributed to the intense growth of population; for instance, Rome, Madrid, Barcelona, Athens, and Naples (including the metropolitan area) had between 2.6 and 3.5 million inhabitants in 1971, with annual growth rates ranging from 2.5 to 3.5% (Fratini 2000; Pacione 2003; Vinci et al. 2022). Another important factor that contributed to the exponential growth of Mediterranean cities during the twentieth century was immigration (Barata Salgueiro 2006). Most of the region experienced intense immigration flows from Northern Africa and Eastern Europe (Nickayin et al.

2020). This phenomenon has had (and is still having) relevant consequences in the social organization of those cities.

Along the last century, a higher demand and desirability for downtown dwelling units was observed in Mediterranean cities, in contrast with the dominant Northern/Western/Central European urban trends (Ondono et al. 2021). Dividing cities in a set of concentric circles expanding from the suburbs to downtown and influencing most socioeconomic traits (Robson 1969), an 'inverse-Burgess' model was observed in various metropolitan regions throughout Southern Europe, possibly resembling the distribution (and re-distribution over time) of social classes according to land-rent and building density gradients (Orenstein et al. 2014; Salvati et al. 2019; Rubiera-Morollòn and Garrido-Yserte 2020). As a consequence, the rich bourgeoisie clustered in central cities, where land-rent values and population density are high, while the working class moved toward peripheral areas of the city where the costs of life are affordable. For instance, social rank in Rome has been found to be low in 'suburbs' (and still lower along the rural fringe, the so-called 'Agro Romano' district) and particularly high downtown (e.g. Salvati al. 2016).

The inverse-Burgess tendency is one of the main reasons that brought to the high frequency of gentrification processes in Mediterranean cities (Storper and Scott 2016). As desirability for central areas grows, many of these districts have undergone intense re-qualification programs (Swyngedouw 2004). Central historical slums, typical of old towns like the Mediterranean ones, have been strongly re-qualified and thus risen in value (Krumholtz 1992). Some examples of this process can be observed in the central districts of Rome, Lisbon, and Barcelona (Zambon et al. 2017). However, even if the inverse-Burgess model is the dominant trend in Mediterranean cities, this does not mean that urbanization patterns always follow it precisely (Fig. 4.9).

Additionally, a sort of 'spatial separation' between the middle and working class had frequently observed downtown, but not specifically along the 'horizontal' axis (Maloutas 2007). As a matter of fact, the middle and working class usually lived together in vertically stratified apartment blocks downtown (Garcia-Lopez and Muniz 2010). Service laborers and working class occupied lower floors, and wealthier population maintained on top levels in penthouses (Wynn 1984; White 1984: 156). This social separation on height levels has been called 'vertical differentiation', as initially observed in the seminal work by Allum (1973) for Naples' downtown. Vertical differentiation has been frequently attributed to the proliferation of multistory apartment buildings (Fig. 4.10).

The location of the economic activities provides us with another contrast between Northern Europe and the Mediterranean spatial patterns of urbanization (Galster 2001; Paul and Tonts 2005; Domene and Sauri 2006; Gonzales 2011). The late industrialization in the Mediterranean region outlines the lack of a 'bourgeoisie hegemony', but above all of heterogeneity and diversity (Saizen et al. 2006; Koomen 2007; Turok and Mykhnenko 2007). Zoning was unfrequently applied in Mediterranean cities (Amato et al. 2015), while mixed land-use seems to be the general trend (Fig. 4.11). By contrast, a polarized organization of business activities in specialized

Fig. 4.9 Time lapse delineating the expansion of metropolitan Rome over one century (**a** 1921, **b** 1971, **c** 2021). *Source* Author's elaboration on digital cartography of census enumeration districts in Italy (building density per hectare, using the same color palette and values' ramp) released by the National Statistical Institute (Istat)

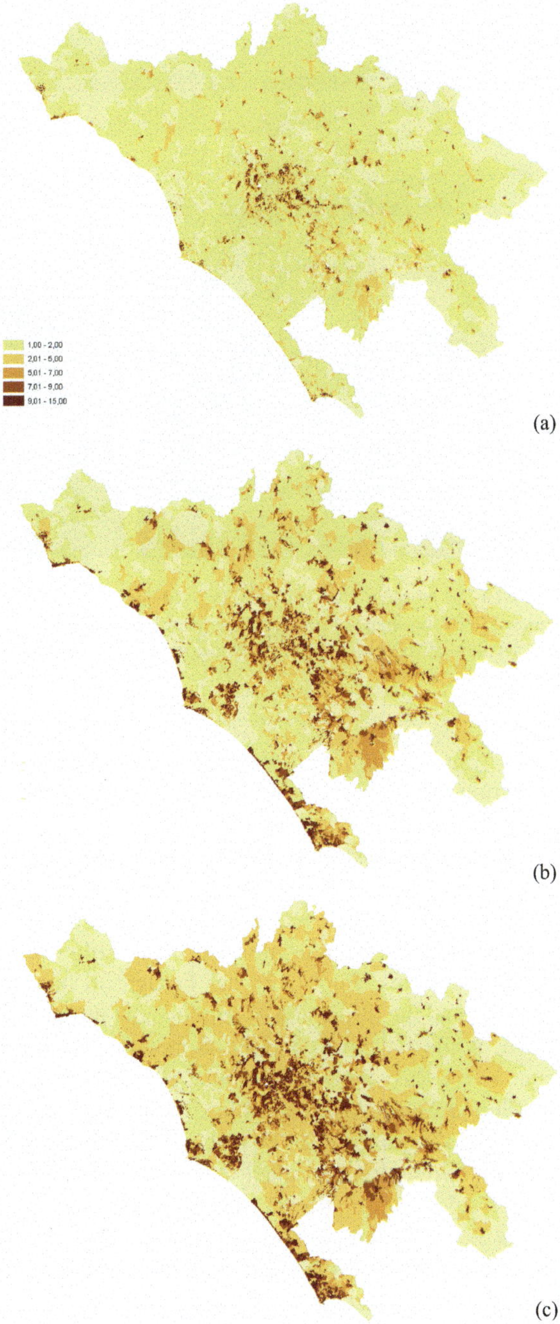

Fig. 4.10 Welcome to the 'vertical city': old and new experiences of compactness and density in Athens; (left) the shape of downtown residential buildings close to Omonia Square, and (right) the new business district of Syngrou Avenue, in-between Kallithea and Nea Smyrni. *Source* Author's photographic archive

zones located according to the land-rent gradient was the result of economic rationalization and planning in Western Europe (Klosterman and Musterd 2001; Longhi and Musolesi 2007; Munoz 2010).

The intimate tissue of Mediterranean cities revealed to be mostly chaotic and heterogeneous (Vinci et al. 2023), with housing areas hosting multiple activities (Wallerstein 1979), and vertical differentiation being the norm (Zambon et al. 2019). All this, together with an informal organization of labor activities and inefficient planning systems (Fig. 4.12), has brought to intermixed working and residential activities and the dispersion of employment throughout the entire city (Zwiers et al. 2016). While effective controls stopped land invasion in Western European cities from a very early period (Basnou et al. 2013), popular land control was out of the question in Anglo-American cities. At the same time, semi-squatting was a diffused practice (Ioannidis et al. 2009) that gave birth to shanty towns in Mediterranean large cities, the strength of so-called 'informal' housing being dominant (Sauri 2003).

In this perspective, Mediterranean large agglomerations were thus built bit by bit (Fig. 4.12), and their physical expansion was defined by where people erected their precarious settlements (Salvati et al. 2018). Italian cities, besides some exceptions in the North of the country, grew mainly through popular land colonization (Salvati and Carlucci 2016). For instance, Rome evidenced a clear pattern of spontaneous development in the 1970s (Salvati 2014). At the same time, Spanish and Portuguese large cities were surrounded by 'viviendas marginales' (Dawkins and Nelson 2002).

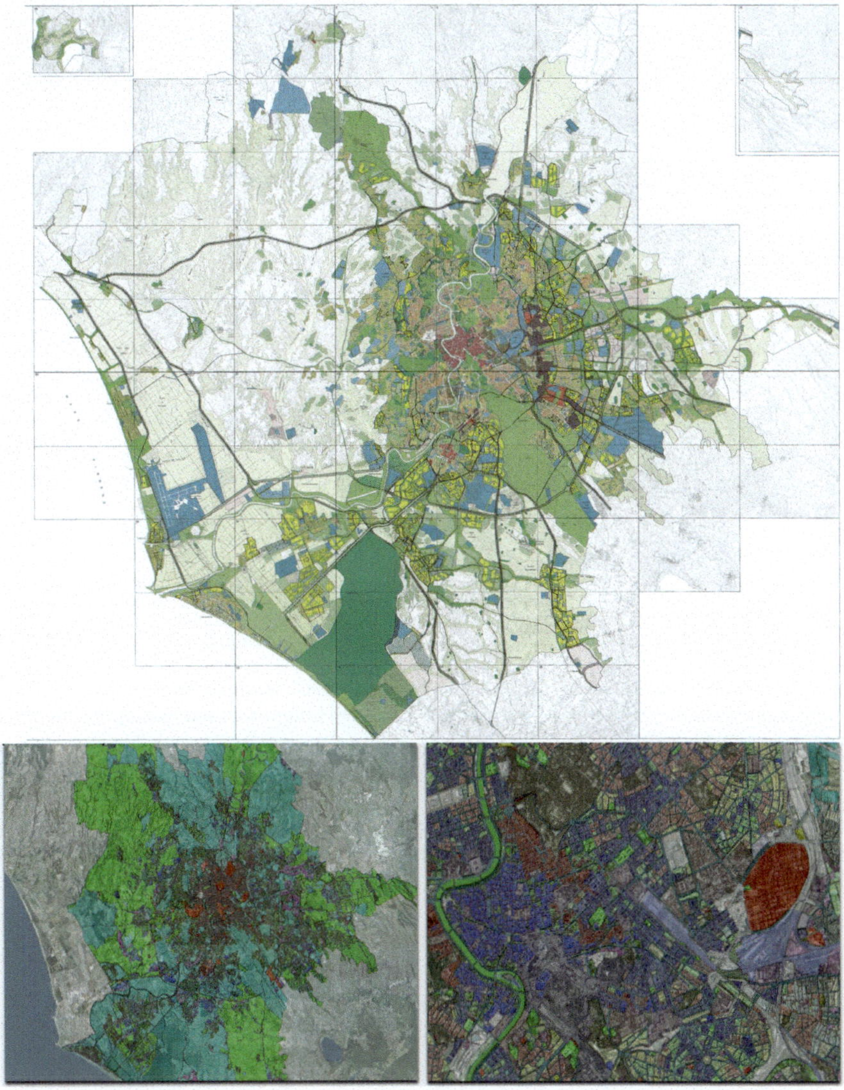

Fig. 4.11 'Uneffable zoning' of Rome from World War II (above: the town master plan dated 1962) to nowadays (below: the new master plan dated 2008 and a focus on the historical city within the Aurelian Walls); sources: author's elaboration on historical archives and geo-spatial databases released by Cartographic Office of Rome province

Fig. 4.12 Mirable compactness of Western Athens' settlements, expanding radio-centrically from the inner city in the 1960s and the 1970s, in a shot taken from Lofos Strefi and looking to Petroupoli on the background. *Source* Author's photographic archive

In the case of Athens, colonization of peripheral zones by the proletariat and floods of migrants to central reception areas has been observed during the post-war period (Alexandri 2015).

4.5 Summarizing Cultural Paths of Metropolitan Growth

Sequential urbanization and suburbanization waves express, likely better than any other economic process, the joint cultural and intellectual characteristics of any human society (Mills 1967). In this perspective, rejecting a unique model of Mediterranean urbanization does not imply that there are not common characteristics that metropolitan centers of this region may share (Abrams 1964; Westerink et al. 2013; Cividino et al. 2020). Urban histories of the Mediterranean feature a glorious past and a period of late industrialization combined with a demographic boom and immigration waves in the twentieth century (Torrens 2006; Verburg et al., 2004; Herrero-Jauregui and Conception 2023). During this period, a higher desirability for city centers and the lack of an organized industrial spatial asset have brought to the concentration of high and middle classes in core cities (Maloutas 2004; Weng 2007;

Arapoglou and Sayas 2009). The economic system is a crucial element shaping the urban asset of a given area, together with a more latent factor of 'popular land control' (sensu Amato 1970), being a dominant trend in specific areas surrounding city centers (Fig. 4.13). However, explaining the spatial distribution of social groups is not sufficient for a complete analysis of 'Southern urbanities' (Karadimitriou et al. 2021), that urgently requires an intimate comprehension of the relationship between social groups and economic assets at the metropolitan scale (Di Feliciantonio et al. 2018).

A coherent analysis of social formation and production modes in Mediterranean communities has clarified how subordinates classes have been dominant in transforming urban landscapes (Kaika 2003), acquiring importance thanks to a pervasive informal economy and weak industrialization processes (Allen et al. 2004; Mukhopadhyay et al. 2020; Randolph and Storper 2023). The notion of 'semi-peripheral cities'—possibly associated with a number of medium-size cities in Southern Europe—refers to the fact that urbanization has followed industrialization processes, in contrast with what has been observed in Northern Europe (Fig. 4.14). Anglo-American and Mediterranean societies were definitely proposed as contrasting examples of a strong anti-urban ideology against a pro-urban ideology, with different implications of urban sustainability at large (Couch et al. 2007; Zhang 2008; Feng et al. 2019; Berisha and Cotella 2021). Our work suggests how pro-urban ideologies

Fig. 4.13 Waiting for afternoon rest: popular land control and the 'green panorama' from Attiko Alsos, Central Athens. *Source* Author's photographic archive

Fig. 4.14 Informal design of industrial settlements from a 'semi-peripheral city', in-between spontaneity and planning; a shot representing the industrial area between Likovrisi and Metamorfosi, Northern Athens. *Source* Authors' own photographic archive

typical of Mediterranean regions should be considered in historical and geographical perspectives when exploring the spatial forms of cities and the inherent linkage with sustainability (Mitchell and Devisscher 2022). Understanding the urban ideology of a society is crucial for determining its expansion (Lindstrom and Bartling 2003; Keddie and Tonkiss 2010; Mouratidis 2019).

4.6 Concluding Remarks

While metropolitan landscapes and land-use patterns have been closely related to economic growth, demographic change, and social dynamics, heterogeneous patterns and processes of metropolization reflect the spatial distribution of agglomeration advantages, the subtle impact of production scale, and the consequent implications for land resource management and the preservation of relict environmental qualities. By defeating the historical pre-existences and the traditional characteristics of the territory, planning deregulation plays a role in this complex interaction, since it tends to irreversibly damage rural landscapes, making settlement forms more and more confused (Richardson and Chang-Hee 2004; Overmars et al. 2007; Sahana et al. 2023).

In a progressive stratification of functions that reflect land speculation, reconnecting regional-scale impacts of economic change with local-scale ecological dynamics contributes to an enriched knowledge of territorial trajectories and sociodemographic histories, outlining how a study of differences under assumptions of nonlinearity and complex system thinking is key to understand future urban trends in the old continent, and especially in Mediterranean Europe. Spatial planning and sustainable land management tactics have, together, delineated novel opportunities to local development, possibly containing the negative impact of 'one-way', 'centralized', and 'top-down' growth processes based on production scale and economic agglomeration (Tsai 2005; Salvati et al. 2012; Scott and Storper 2015).

In a strategy oriented toward sustainable development, destining precious urban (and peri-urban) land to privileged activities such as research, tourism, services, and craftsmanship, quality agri-food production should be combined with various forms of incentives/subsidies to settlement containment, professional training, and facilitation of 'brownfield' development. Recovering urban depressed areas from an economic and social point of view means reducing the impact of 'land speculation', and avoiding the well-known outcomes of sprawl characteristic of decomposed forms and uncoordinated, random functions (Salvati 2024).

With this perspective in mind, the empirical results of our study encourage multidisciplinary research and stimulate further studies that concentrate on the deserving issue of sustainability—more tightly interconnected with urban cultures and the social sphere. Although the substantial contribution of narrative and visual approaches is evident from this chapter, quantitative information—possibly based on official statistics and elementary indicators —is increasingly needed to reach the desired level of completeness and reliability requested in policy implementation. Policy strategies grounded on concrete, practical actions going beyond the constraint of the geographical scale and constructing an alternative to the 'lassez-faire planning', are increasingly requested in the Mediterranean countries.

New theoretical frameworks, together with empirical approaches, comparative works, and case studies, have to shed additional light in the intrinsic nexus between individual paths of city growth. Together with such research advancements, a better knowledge of urban cultures—and metropolitan sustainability at large—will provide an appropriate and informed ground to science and policy.

References

Abrams C (1964) Man's struggle for shelter in an urbanizing world. MIT Press, Cambridge, Mass

Adger WN, Crépin AS, Folke C, Ospina D, Chapin FS, Segerson K, Wilen J (2020) Urbanization, migration, and adaptation to climate change. One Earth 3(4):396–399

Alexandri G (2015) Reading between the lines: gentrification tendencies and issues of urban fear in the midst of Athens' crisis. Urban Stud 52(9):1631–1646

Allen J, Barlow J, Maloutas T, Padovan L (2004) Housing in southern Europe. Blackwell, London

Allum P (1973) Politics and society in post-war Naples. Cambridge University Press, Cambridge
Amato P (1970) A comparison: population densities, land values and socio-economic class in four Latin American cities. Land Econ 46:447–451
Amato F, Pontrandolfi P, Murgante B (2015) Supporting planning activities with the assessment and the prediction of urban sprawl using spatio-temporal analysis. Ecol. Inf. 30:365–378
Andreotti A, Le Galès P, Moreno Fuentes FJ (2013) Controlling the urban fabric: the complex game of distance and proximity in European upper-middle-class residential strategies. Int J Urban Reg Res 37(2):576–597
Angel S, Parent J, Civco DL, Blei A, Potere D (2011) The dimensions of global urban expansion: estimates and projections for all countries, 2000–2050. Prog Plan 75(2):53–107
Arapoglou VP, Sayas J (2009) New facets of urban segregation in Southern Europe. Eur Urban Reg Stud 16(4):345–362
Arbaci S (2008) (Re)Viewing ethnic residential segregation in southern European cities: housing and urban regimes as mechanisms of marginalisation. Hous Stud 23(4):589–613
Arbaci S, Malheiros J (2010) De-segregation, peripheralisation and the social exclusion of immigrants: southern European cities in the 1990s. J Ethn Migr Stud 36(2):227–255
Balchin P (1996) Housing policy in Europe. Routledge, London and New York
Barata Salgueiro T (2006) Lisbonne. Périphérie et centralités. L'Harmattan, Paris
Basnou C, Álvarez E, Bagaria G, Guardiola M, Isern R, Vicente P, Pino J (2013) Spatial patterns of land use changes across a Mediterranean metropolitan landscape: implications for biodiversity management. Environ Manage 52:971–980
Berisha E, Cotella G (2021) Towards sustainable urbanization. Learn from what's out there. Land 10(4):356
Bialasiewicz L, Minca C (2005) Old Europe, new Europe: for a geopolitics of translation. Area 37(4):365–372
Bialasiewicz L, Giaccaria P, Jones A, Minca C (2013) Re-scaling 'EU'rope: EU macro-regional fantasies in the Mediterranean. Eur Urban Reg Stud 20(1):59–76
Bianchini L, Egidi G, Alhuseen A, Sateriano A, Cividino S, Clemente M, Imbrenda V (2021) Toward a dualistic growth? Population increase and land-use change in Rome. Italy. Land 10(7):749
Braudel F (1953) Civiltà e Imperi del Mediterraneo nell'età di Filippo II. Einaudi, Torino
Braudel F (1987) Il Mediterraneo—Lo Spazio la Storia gli Uomini le Tradizioni. Bompiani, Milano
Bruegmann R (2019) Sprawl: a compact history. University of Chicago Press, Chicago
Bueno-Suárez C, Coq-Huelva D (2020) Sustaining what is unsustainable: a review of urban sprawl and urban socio-environmental policies in North America and Western Europe. Sustainability 12(11):4445
Buzar S, Ogden P, Hall R, Haase A, Kabisch S, Steinfiihrer A (2007) Splintering urban populations: emergent landscapes of reurbanisation in four European cities. Urban Stud 44(4):651–677
Cai Z, Liu Q, Cao S (2020) Real estate supports rapid development of China's urbanization. Land Use Policy 95:104582
Cassiers T, Kesteloot C (2012) Socio-spatial inequalities and social cohesion in European cities. Urban Stud 49(9):1909–1924
Castells-Quintana D, Wenban-Smith H (2020) Population dynamics, urbanisation without growth, and the rise of megacities. J Dev Stud 56(9):1663–1682
Cazcarro I, Duarte R, Sánchez-Chóliz J (2012) Economic growth and the evolution of water consumption in Spain: a structural decomposition analysis. Ecol Econ 96:51–61
Chorianopoulos I, Pagonis T, Koukoulas S, Drymoniti S (2010) Planning, competitiveness and sprawl in the Mediterranean city: the case of Athens. Cities 27:249–259
Chorianopoulos I, Tsilimigkas G, Koukoulas S, Balatsos T (2014) The shift to competitiveness and a new phase of sprawl in the Mediterranean city: enterprises guiding growth in Messoghia-Athens. Cities 39:133–143
Cividino S, Halbac-Cotoara-Zamfir R, Salvati L (2020) Revisiting the "city life cycle": global urbanization and implications for regional development. Sustainability 12(3):1151

Costa FJ, Noble AG, Pendeleton G (1991) Evolving planning systems in Madrid, Rome, and Athens. GeoJournal 24(3):293–303

Couch C, Petschel-Held G, Leontidou L (2007) Urban sprawl in Europe: landscapes, land-use change and policy. Blackwell, Oxford

Cuadrado-Ciuraneta S, Durà-Guimerà A, Salvati L (2017) Not only tourism: Unravelling suburbanization, second-home expansion and "rural" sprawl in Catalonia, Spain. Urban Geogr 38(1):66–89

Dawkins CJ, Nelson AC (2002) Urban containment policies and housing prices: an international comparison with implications for future research. Land Use Policy 19:1–12

De Rosa S, Salvati L (2016) Beyond a 'side street story'? Naples from spontaneous centrality to entropic polycentricism, towards a 'crisis city.' Cities 51:74–83

De Muro P, Monni S, Tridico P (2011) Knowledge-Based Economy And Social Exclusion: Shadow And Light In the Roman socio-economic model. Int J Urban Reg Res 35(6):1212–1238

Di Feliciantonio C, Salvati L (2015) 'Southern' alternatives of urban diffusion: investigating settlement characteristics and socio-economic patterns in three Mediterranean regions. Tijdschr Econ Soc Geogr 106(4):453–470

Di Feliciantonio C, Salvati L, Sarantakou E, Rontos K (2018) Class diversification, economic growth and urban sprawl: evidences from a pre-crisis European city. Qual Quant 52:1501–1522

Domene E, Sauri D (2006) Urbanization and water consumption: influential factors in the metropolitan region of Barcelona. Urban Stud 43(9):1605–1623

Domene E, Sauri D, Parés M (2005) Urbanization and sustainable resource use: the case of garden watering in the metropolitan region of Barcelona. Urban Geogr 26(6):520–535

Durà-Guimera A (2003) Population deconcentration and social restructuring in Barcelona, a European Mediterranean city. Cities 20(6):387–394

Feng Y, Wang X, Du W, Liu J, Li Y (2019) Spatiotemporal characteristics and driving forces of urban sprawl in China during 2003–2017. J Clean Prod 241:118061

Ferras R (1977) Barcelone: Croissance d'une Metropole. Anthropos, Paris

Fratini F (2000) Roma arcipelago di isole urbane. Gangemi, Rome

Fried RC (1973) Planning the eternal city: Roman politics and planning since World War II. Yale University Press, London

Galster G, Hanson R, Ratcliffe MR, Wolman H, Coleman S, Freihage J (2001) Wrestling sprawl to the ground: defining and measuring an elusive concept. Hous Policy Debate 12(4):681–717

Gao M, Chen B, Xu Y, Li D (2023) Heterogeneous impacts of global land urbanization on land-use structure from economic and technological perspectives. Ecol Ind 147:109955

Garcia-López MÀ (2019) All roads lead to Rome… and to sprawl? Evidence from European cities. Reg Sci Urban Econ 79:103467

Garcia-López MA, Muñiz I (2010) Employment decentralisation: polycentricity or scatteration? The case of Barcelona. Urban Stud 47(14):3035–3056

Giaccaria P, Minca C (2011) The mediterranean alternative. Prog Hum Geogr 35(3):345–365

Glaeser EL (2020) Urbanization and its discontents. East Econ J 46(2):191–218

Gonzales S (2011) The north/south divide in Italy and England: discursive construction of regional inequality. Eur Urban Reg Stud 18(1):62–76

Gospodini A (2009) Post-industrial trajectories of Mediterranean European cities: the case of post-Olympics Athens. Urban Stud 46(5–6):1157–1186

Gross J, Ouyang Y (2021) Types of urbanization and economic growth. Int J Urban Sci 25(1):71–85

Grout C, Cavailhès J, Détang-Dessendre C, Thomas A (2016) Is Sprawling residential behavior influenced by climate? Land Econ 92(2):203–219

Hall P (1997) The future of metropolis and its form. Reg Stud 31(3):211–220

Hamnett C (2020) Is Chinese urbanisation unique? Urban Stud 57(3):690–700

Handlin O (1963) The historian and the city. Cambridge University Press, Cambridge

Herrero-Jáuregui C, Concepción ED (2023) Effects of counter-urbanization on Mediterranean rural landscapes. Landsc Ecol 38(12):3695–3711

Ioannidis C, Psaltis C, Potsiou C (2009) Towards a strategy for control of suburban informal buildings through automatic change detection. Comput Environ Urban Syst 33:64–74
Ismael HM (2021) Urban form study: the sprawling city—review of methods of studying urban sprawl. GeoJournal 86(4):1785–1796
Kaika M (2003) Constructing scarcity and sensationalising water politics: 170 days that shook Athens. Antipode 35(5):919–954
Karadimitriou N, Maloutas T, Arapoglou VP (2021) Multiple deprivation and urban development in Athens, Greece: spatial trends and the role of access to housing. Land 10(3):290
Keddie J, Tonkiss F (2010) The market and the plan: housing, urban renewal and socio-economic change in London. City Cult Soc 1(2):57–67
Kemeny T, Storper M (2012) The sources of urban development: wage, housing, and amenity gaps across American cities. J Reg Sci 52(1):85–108
Khan I, Hou F, Le HP, Ali SA (2021) Do natural resources, urbanization, and value-adding manufacturing affect environmental quality? Evidence from the top ten manufacturing countries. Resour Policy 72:102109
King R, Proudfoot L, Smith B (1997) The Mediterranean. Environ Soc
Kloosterman RC, Musterd S (2001) The polycentric urban region: towards a research agenda. Urban Stud 38(4):623–633
Koomen E (2007) Modelling land-use change: progress and applications. Springer
Koprowska K, Łaszkiewicz E, Kronenberg J (2020) Is urban sprawl linked to green space availability? Ecol Ind 108:105723
Kroll F, Kabisch N (2012) The relation of diverging urban growth processes and demographic change along an urban-rural gradient. Popul Space Place 18:260–276
Krumholz N (1992) Roman impressions: contemporary city planning and housing in Rome. Landsc Urban Plan 22(2–4):107–114
Kundu D, Sharma P, Gupta D, Mueller A, Gareis P (2023) Measuring sustainable urban development: the case of India, Europe and Germany. Environ Urban ASIA 14(2):269–284
Le Goix R (2005) Gated communities: sprawl and social segregation in southern California. Hous Stud 20(2):323–343
Lee S (2011) Metropolitan growth patterns and socio-economic disparity in six US metropolitan areas 1970–2000. Int J Urban Reg Res 35(5):988–1011
Leontidou L (1990) The Mediterranean city in transition—social change and urban development. Cambridge University Press, New York
Leontidou L (1996) Mediterranean cities: divergent trends in a United Europe. In: Blacksell et al (eds)The European challenge: geography and development in the European Community. Oxford University Press, Oxford, pp 127–148
Lindstrom MJ, Bartling H (2003) Suburban sprawl: culture, theory, and politics. Rowman & Littlefield, New York
Liu L, Meng L (2020) Patterns of urban sprawl from a global perspective. J Urban Plann Dev 146(2):04020004
Longhi C, Musolesi A (2007) European cities in the process of economic integration: towards structural convergence. Ann Reg Sci 41(2):333–351
Maloutas T (2007) Socio-economic classification models and contextual difference: the "European socio-economic classes" (ESeC) from a South European angle. South Eur Soc Polit 12(4):443–460
Maloutas T (2004). Segregation and residential mobility. Socially entrapped social mobility and its impact on segregation in Athens. Eur Urban Reg Stud 11(2):171–187
Masini E, Tomao A, Barbati A, Corona P, Serra P, Salvati L (2019) Urban growth, land-use efficiency and local socioeconomic context: a comparative analysis of 417 metropolitan regions in Europe. Environ Manage 63:322–337
Menashe-Oren A, Bocquier P (2021) Urbanization is no longer driven by migration in low-and middle-income countries (1985–2015). Popul Dev Rev 47(3):639–663

Mills ES (1967) An aggregative model of resource allocation in a metropolitan area. Am Econ Rev 57:197–210
Mitchell MG, Devisscher T (2022) Strong relationships between urbanization, landscape structure, and ecosystem service multifunctionality in urban forest fragments. Landsc Urban Plan 228:104548
Mouratidis K (2019) Compact city, urban sprawl, and subjective well-being. Cities 92:261–272
Mudu P (2006) Patterns of segregation in contemporary Rome. Urban Geogr 27(5):422–440
Mukhopadhyay P, Zérah MH, Denis E (2020) Subaltern urbanization: Indian insights for urban theory. Int J Urban Reg Res 44(4):582–598
Muñoz F (2003) Lock living: urban sprawl in Mediterranean cities. Cities 20(6):381–385
Muñoz F (2010) Urbanalization: common landscapes, global places. Open Urban Stud J 3:78–88
Murtin F (2013) Long-term determinants of the demographic transition, 1870–2000. Rev Econ Stat 95(2):617–631
Musterd S (2005) Social and ethnic segregation in Europe: levels, causes and effects. J Urban Aff 27(3):331–348
Musterd S, Marcińczak S, Van Ham M, Tammaru T (2017) Socioeconomic segregation in European capital cities. Increasing separation between poor and rich. Urban Geogr 38(7):1062–1083
Newman P, Thornely A (1996) Urban planning in Europe. Routledge, New York
Nickayin SS, Tomao A, Quaranta G, Salvati L, Gimenez Morera A (2020) Going toward resilience? Town planning, peri-urban landscapes, and the expansion of Athens, Greece. Sustainability 12(24):10471
Oddo L, Zanini A (2022) The paradox of "Malthusian urbanization": urbanization without growth in the Republic of Genoa, 1300–1800. Eur Rev Econ Hist 26(4):508–534
Ondoño IS, Abellán FC, Garcia-Gonzalez JA (2021) The Cadastre as a source for the analysis of urbanization dynamics. Applications in urban areas of medium-sized inland Spanish cities. Land 10(4):1–23
Orenstein DE, Frenkel A, Jahshan F (2014) Methodology matters: measuring urban spatial development using alternative methods. Environ Plann B Plann des 41:3–23
Overmars KP, Verburg PH, Veldkamp A (2007) Comparison of a deductive and an inductive approach to specify land suitability in a spatially explicit land use model. Land Use Policy 24:584–599
Pacione M (2003) Urban environmental quality and human wellbeing - a social geographical perspective. Landsc Urban Plan 65:19–30
Pandey B, Brelsford C, Seto KC (2022) Infrastructure inequality is a characteristic of urbanization. Proc Natl Acad Sci 119(15):e2119890119
Parés M, March H, Saurí D (2013) Atlantic gardens in Mediterranean climates: understanding the production of suburban natures in Barcelona. Int J Urban Reg Res 37:328–347
Paul V, Tonts M (2005) Containing urban sprawl: trends in land-use and spatial planning in the Metropolitan Region of Barcelona. J Environ Planning Manage 48(1):7–35
Power A (2001) Social Exclusion and Urban Sprawl: is the rescue of cities possible? Reg Stud 35(8):731–742
Profiroiu CM, Bodislav DA, Burlacu S, Rădulescu CV (2020) Challenges of sustainable urban development in the context of population Growth. Eur J Sustain Dev 9(3):51–51
Randolph GF, Storper M (2023) Is urbanisation in the Global South fundamentally different? Comparative global urban analysis for the 21st century. Urban Stud 60(1):3–25
Richardson H, Chang-Hee C (2004) Urban sprawl in Western Europe and the United States. Ashgate, London
Robson BT (1969) Urban analysis. Cambridge University Press, Cambridge
Rontos K, Grigoriadis S, Sateriano A, Syrmali M, Vavouras I, Salvati L (2016) Lost in protest, found in segregation: divided cities in the light of the 2015 'Oki' Referendum in Greece. City Cult Soc 7(3):139–148

Rubiera-Morollón F, Garrido-Yserte R (2020) Recent literature about urban sprawl: a renewed relevance of the phenomenon from the perspective of environmental sustainability. Sustainability 12(16):6551
Sahana M, Ravetz J, Patel PP, Dadashpoor H, Follmann A (2023) Where is the peri-urban? A systematic review of peri-urban research and approaches for its identification and demarcation worldwide. Remote Sens 15(5):1316
Saizen I, Mizuno K, Kobayashi S (2006) Effects of land-use master plans in the metropolitan fringe of Japan. Landsc Urban Plan 78:411–421
Salvati L (2014) Towards a polycentric region? The socioeconomic trajectory of Rome, an 'eternal' Mediterranean city. Tijdschr Econ Soc Geogr 105(3):268–284
Salvati L (2024) Simple geographies for complex problems? Revisiting long-term urbanization and settlement population mismatches using elasticity indicators and context-based nonparametric analysis. Ann Reg Sci 72(2):513–533
Salvati L, Carlucci M (2016) Patterns of sprawl: the socioeconomic and territorial profile of dispersed urban areas in Italy. Reg Stud 50(8):1346–1359
Salvati L, Munafo M, Morelli VG, Sabbi A (2012) Low-density settlements and land use changes in a Mediterranean urban region. Landsc Urban Plan 105:43–52
Salvati L, Morelli VG, Rontos K, Sabbi A (2013) Latent exurban development: city expansion along the rural-to-urban gradient in growing and declining regions of southern Europe. Urban Geogr 34(3):376–394
Salvati L, Ridolfi E, Pujol DS, Ruiz PS (2016) Latent sprawl, divided Mediterranean landscapes: urban growth, swimming pools, and the socio-spatial structure of Athens, Greece. Urban Geogr 37(2):296–312
Salvati L, Ferrara A, Chelli F (2018) Long-term growth and metropolitan spatial structures: an analysis of factors influencing urban patch size under different economic cycles. Geografisk Tidsskrift—Dan J Geogr 118(1):56–71
Salvati L, Ciommi MT, Serra P, Chelli FM (2019) Exploring the spatial structure of housing prices under economic expansion and stagnation: the role of socio-demographic factors in metropolitan Rome, Italy. Land Use Policy 81:143–152
Sassen S (2010) The city: its return as a lens for social theory. City Cult Soc 1(1):3–11
Saurí D (2003) Lights and shadows of urban water demand management. The case of the metropolitan region of Barcelona. Eur Plann Stud 11(3):229–243
Scott AJ, Storper M (2015) The nature of cities: the scope and limits of urban theory. Int J Urban Reg Res 39(1):1–15
Seronde Babonaux A (1983) Roma. Dalla città alla metropoli. Editori Riuniti, Rome
Shaw BJ, van Vliet J, Verburg PH (2020) The peri-urbanization of Europe: a systematic review of a multifaceted process. Landsc Urban Plan 196:103733
Spina MC (2004) Le città del Mediterraneo. La Biblioteca del Mediterraneo, Perugia
Storper M, Scott AJ (2016) Current debate in urban theory: a critical assessment. Urban Stud 53(6):1114–1136
Sumari NS, Cobbinah PB, Ujoh F, Xu G (2020) On the absurdity of rapid urbanization: spatio-temporal analysis of land-use changes in Morogoro, Tanzania. Cities 107:102876
Sun L, Chen J, Li Q, Huang D (2020a) Dramatic uneven urbanization of large cities throughout the world in recent decades. Nat Commun 11(1):5366
Sun M, Wang J, He K (2020b) Analysis on the urban land resources carrying capacity during urbanization—a case study of Chinese YRD. Appl Geogr 116:102170
Swyngedouw E (2004) Social power and the urbanization of water. Oxford University Press, Oxford, UK
Tian Y, Tsendbazar NE, van Leeuwen E, Fensholt R, Herold M (2022) A global analysis of multi-faceted urbanization patterns using Earth observation data from 1975 to 2015. Landsc Urban Plan 219:104316
Torrens PM (2006) Simulating sprawl. Ann Assoc Am Geogr 96(2):248–275

Townsend A (2009) Integration of economic and spatial planning across scales. Int J Public Sect Manag 22(7):643–659

Tsai Y (2005) Quantifying urban form: compactness versus sprawl. Urban Studies 42(1):141–161

Tsilimigkas G, Kizos T (2014) Space, pressures and the management of the Greek landscape. Geografiska Annaler Ser B Human Geogr 96(2):159–175

Turok I, Mykhnenko V (2007) The trajectories of European cities, 1960–2005. Cities 24:165–182

Verburg PH, Schot PP, Dijst MJ, Veldkamp A (2004) Land use change modelling: current practice and research priorities. GeoJournal 61(4):309–324

Vidal M, Domene E, Saurí D (2011) Changing geographies of water-related consumption: residential swimming pools in suburban Barcelona. Area 43(1):67–75

Vinci S, Egidi G, López Gay A, Salvati L (2022) Population growth and urban management in metropolitan regions: the contribution of natural balance and migration to polycentric development in Barcelona. Appl Spat Anal Policy 15(1):71–94

Vinci S, Vardopoulos I, Salvati L (2023) A tale of a shrinking City? Exploring the complex interplay of socio-demographic dynamics in the recent development of Attica, Greece. Cities 132:104089

Wallerstein I (1979) The capitalist world-economy. Cambridge University Press, Cambridge

Wang Q, Lin J, Zhou K, Fan J, Kwan MP (2020) Does urbanization lead to less residential energy consumption? A comparative study of 136 countries. Energy 202:117765

Weng YC (2007) Spatiotemporal changes of landscape pattern in response to urbanization. Landsc Urban Plann 81:341–353

Westerink J, Haase D, Bauer A, Ravetz J, Jarrige F, Aalbers CBEM (2013) Dealing with sustainability trade-offs of the compact city in peri-urban planning across European city regions. Eur Plan Stud 21(4):473–497

White P (1984) The West European city: a social geography. Longman, London

Wynn M (1984) Planning and urban growth in Southern Europe. Mansell, London

Xiong W, Chen B, Wang H, Zhu D (2020) Public–private partnerships as a governance response to sustainable urbanization: lessons from China. Habitat Int 95:102095

Zambon I, Serra P, Sauri D, Carlucci M, Salvati L (2017) Beyond the 'Mediterranean city': socioeconomic disparities and urban sprawl in three Southern European cities. Geografiska Annaler Ser B Human Geogr 99(3):319–337

Zambon I, Colantoni A, Salvati L (2019) Horizontal vs vertical growth: understanding latent patterns of urban expansion in large metropolitan regions. Sci Total Environ 654:778–785

Zapata-Barrero R (2023) New scales of migration governance in the Mediterranean: regional cities in the spotlight. Eur Urban Reg Stud 30(2):121–134

Zhang W (2008) A forecast analysis on world population and urbanization process. Environ Dev Sustain 10:717–730

Zhang L, Yang L, Zohner CM, Crowther TW, Li M, Shen F, Zhou C (2022) Direct and indirect impacts of urbanization on vegetation growth across the world's cities. Sci Adv 8(27):eabo0095

Zwiers M, Bolt G, van Ham M, Van Kempen R (2016) The global financial crisis and neighbourhood decline. Urban Geogr 37(5):664–684

Chapter 5
Searching for a New Identity? Sustainable Development, 'Crisis' Landscapes, and the Metropolitan Continuum

Abstract A comprehensive evaluation of urban and suburban landscapes under expansions and recessions requires an in-depth understanding of historical, cultural, and political aspects and implications of long-term economic dynamics and short-term shocks. A tight integration between quantitative methodologies and qualitative approaches seems to be particularly appropriate when reading territorial complexity. By integrating different disciplinary visions, this chapter reflects on the latent evolution of spatially varying (urban and rural) landscapes. In a time of crisis, an increasing debate has been promoted on land quality, considering both agricultural and productive activities along the fringe, and on the related socioeconomic relations within metropolitan continuums. In a sustainable development perspective, urban areas can be seen as multifunctional districts favoring diversified economies, and promoting agriculture-related land-use possibly containing settlement expansion. Going beyond the aesthetic-visual approach typical of landscape studies, and reflecting a regional science perspective, a sustainable management of the intrinsic metropolitan spaces that are not strictly urban—and not strictly rural—is the final objective of our study. With this perspective in mind, an action plan stimulating virtuous paths and processes of sustainable development is finally delineated, revitalizing degraded environments in a time of reduced public budgets, and declining social cohesion in metropolitan regions.

Keywords Urbanization · Social disparities · Territorial actors · Planning · Southern Europe

5.1 Introduction

Complex paths of metropolitan expansion diverging along sequential economic downturns, had demonstrated a variable impacts on multiple dimensions of the city life cycle, including land fragmentation, transport accessibility, place attractiveness, and natural amenities (Oueslati et al. 2015). For a long time, economic shocks have influenced metropolitan growth, altering building cycles and shaping house and labor

L. Salvati, *Economic Growth, Regional Disparities, and Urban Sprawl*, Springer Geography, https://doi.org/10.1007/978-3-031-85799-7_5

markets as a result of rising social disparities, and a spatially polarized distribution of businesses (Allen et al. 2004; Barras 2009; Andreotti et al. 2013). In the last decades, heterogeneous models of metropolitan expansion and economic growth were observed in the old continent, becoming a matter of concern for spatial planning, and claiming for specific development policies (Couch et al. 2007; Cebrian Abellan and Sanchez Ondono 2019; Robert et al. 2019).

Continental Europe, as other economically advanced world regions (Angel et al. 2011), has experienced population growth, industrial development, and a compact-dense expansion of cities in the decades following World War II (European Environment Agency 2006). Since the early 1980s, however, more evident spatial patterns toward exurban development have been recorded (Batty and Xie 1994; Brade et al. 2009; Aguilera et al. 2011; Burger and Meijers 2012). In this context, the 2007 recession in the old continent was typically originated in urban spaces, and then became a 'state crisis' (Morelli et al. 2014; Alexandri 2015; Rubiera Morollon et al. 2016), with consequences and impacts for subnational scales—namely regions, industrial districts, central localities, and municipalities (Domene et al. 2005; Cakir et al. 2008; Carlucci et al. 2018).

In times of crisis, understanding the latent evolution of both urban and suburban territories requires a deep reflection on the role of industrial, agricultural, and service activities typical of fringe land, in a perspective of sustainable local development (Darques 2015; Garcia-Lopez 2019; Lagarias and Stratigea 2023). In this sense, rural peripheries were often seen as multifunctional areas based on diversified economies and promoting agriculture-related land use (Economou 1997; Carlucci et al. 2017; Garcia-Nieto et al. 2018). This contribution poses a specific reflection on urban and suburban landscapes under recession, integrating different disciplinary visions (Pozokidou and Ntriankos 2017; Feng et al. 2019; Salvati 2019). Non-urban and, together, non-rural landscapes are a target of this study, going beyond the aesthetic-visual approach typical of landscape/architectural studies (Di Feliciantonio and Salvati 2015).

Reading and interpreting territorial complexity definitely require a tight integration of quantitative methodologies and qualitative approaches (De Rosa and Salvati 2016). Coupled with an extensive literature review, the empirical results of this study outline how a coherent evaluation of latent landscapes under economic crisis demands an in-depth investigation of historical, cultural, and identity values at the base of contemporary socioeconomic systems (De Muro et al. 2011). This approach will also suggest appropriate strategies that can trigger virtuous processes of sustainable development and revitalize potentially degraded environments (Giordano and Roller 2004; Brenner et al. 2012; Knieling 2015).

5.2 Stylized Facts

Economic crisis has intrinsically shaped metropolitan growth, possibly altering building cycles and re-organizing house and labor markets as a consequence of rising disparities in businesses, dwelling quality, and social classes (Goldblum and Wong 2000; Winarso and Firman 2002; Pérez 2010; Ren 2015). At the same time, urban municipalities, cities, and metropolitan regions became (more and more) significant scales of geography at which to examine the economic, political, and social implications of austerity (Arribas-Bel et al. 2011; Cassiers and Kesteloot 2012; Venanzoni et al. 2017). However, empirical evidence on the impact of the most recent economic stagnations on different patterns and processes of metropolitan growth are rather mixed, since they mostly refer to place-specific variables (Fig. 5.1). These driving factors of metropolitan change, social transformations, and economic performances were assessed over relatively short time windows and/or within restricted districts, areas, or regions in a given country (Delladetsima 2006; Garcia 2010; Cho et al. 2014).

In the light of economic (and building) cycles, urbanization and suburbanization have been analyzed from vastly different disciplinary perspectives—from merely morphological to strictly functional approaches, adopting multiple indicators and heterogeneous geographical scales interpreting territorial dynamics and planning prescriptions (Arbaci 2008; Bajocco et al. 2012; Biber and O'Neill 2021). The limit of empirical studies dealing with the impact of economic shocks on metropolitan transformations is in the use of—traditional and advanced—indicators that evaluate specific aspects of buildings, construction industry, real estate strategies, and housing patterns (Cazcarro et al. 2012; Pardo-Garcia and Merida-Rodriguez 2018; Fuladlu et al. 2021). Their availability is usually constrained at high-resolution scales, while temporal update is frequently delayed because of expensive collection procedures, sampling surveys, privacy law rules, and other restrictions of official statistics (e.g. Briassoulis 2005; El Menchawy et al. 2011; Morote and Hernandez 2016).

Empirical evidence reveals how recessions impacted negatively the demand of new dwellings, with (more or less intense) consequences for real estate local markets (Bruegmann 2019). Building activity has consequently declined, driving—at least in some cases—a latent trend toward settlement containment (Arbaci and Malheiros 2010). Based on empirical findings from official statistics and planning indicators, Catalàn et al. (2008) argued that settlement sprawl—taken as an individual path of urban growth over a defined time window—has displayed different traits compared with 'westernized' urban experiences (Florida 2011). This was the result of specific legislative and political conditions, with changes in these contexts mainly driven by economic stagnation and the consequent metropolitan decline (e.g. Bueno-Suarez and Coq-Huelva 2020). For instance, a progressive return to inner cities in Spain materialized with more frequent rehabilitation interventions. These had a strong policy support that coincided with population decline (Galster et al. 2001), in turn associated with an acute industrial crisis, new democratic processes in regional

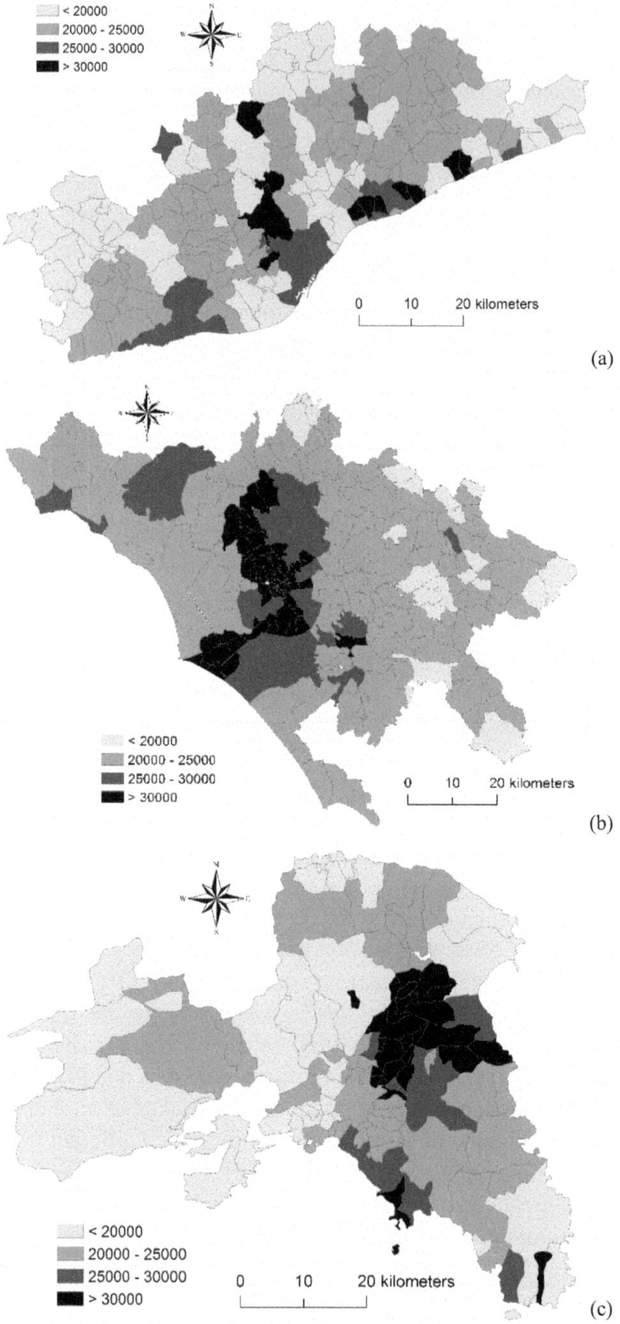

Fig. 5.1 Geography of wealth in three Mediterranean metropolises based on the spatial distribution of per capita income (Euros) in municipalities and local districts of Barcelona (**a**), Rome (**b**), and Athens (**c**) metropolitan regions. *Source* Author's elaboration of official statistics from the respective statistical institute of Spain (INE), Italy (ISTAT) and Greece (ELSTAT)

authorities, and a renewed role of municipal councils in participatory planning of metropolitan spaces (Pérez 2010).

At the same time, the great recession was intended as a basic opportunity for revisiting metropolitan development models toward criteria for both sustainability and resilience (Schneider et al. 2010). Population densification and settlement compactness have enhanced local diversity in land-use and architectonical designs. Re-use of brownfield land and a robust program of building rehabilitation were the possible, indirect, outcomes of recession, in spite of big development programs, land speculation, international real estate investments, and settlement expansion into greenfield (Gospodini 2001, 2006, 2009; Cuadrado-Ciuraneta et al. 2017). Following McCharty (2012),

> the growing recognition of a real estate bubble prompted intense speculation regarding the possibility that the landscape of suburban sprawl might have to be rethought along lines that would emphasise mass transit and greater density.

Under the assumption that settlement morphologies and (land) resource consumption co-evolve dynamically with public finances (Giannakourou 2005), earlier studies have delineated some differences in the rate of population growth, settlement increase, and land-use change between expansions and stagnations, leading to possibly divergent socioeconomic outcomes and territorial backgrounds at the local scale (Domene and Sauri 2006; Gonzales 2011; Di Feliciantonio et al. 2018). From a planning perspective, the seminal study by Cho et al. (2014) suggests how, likely for the first time in the literature, a planned boundary for city growth attracts building projects during expansions (Fig. 5.2), while its effectiveness decreases drastically with recession (Coccossis et al. 2005). At the same time, it was demonstrated how agricultural zoning works better in restraining new development programs during expansions (Cohen 2006). Finally, land-value taxation may increase the incentive for building during recessions, while higher tax bills do not seem to influence building programs during expansions (Ciommi et al. 2019).

From the individual perspective, it was demonstrated how consumers' marginal willingness to pay for landscape attributes decreased systematically with recession, likely because of a progressive deterioration in the economic condition of households (Garcia-Lopez and Muniz 2010; Chorianopoulos et al. 2014; Ciommi et al. 2018). During an economic boom, a strong preference of developers for cheaper land was sometimes documented (Grout et al. 2016). Agricultural conversion to urban use thus realized farther away from urban and even suburban nodes (Emmanuel 2014); at the same time, while remaining substantially inconsistent with actual land-use distribution over small areas, zoning was still considered a proxy of future development trends (Dura-Guimera 2003). Based on this interpretative framework, development paths modified drastically during economic busts (Chorianopoulos et al. 2010).

This path disproportionately leveraged agricultural conversions—especially in fast-growing districts—highlighting the impact financial environments exert on land-use transformations (Delfanti et al. 2016; Greenberg 2021; Inglis et al. 2023). By comparing pre-crisis with post-crisis dynamics, changes over time in the spatial distribution of resident population were considered a local-scale indicator of recent growth

Fig. 5.2 Contested landscapes along Malaga's fringe. *Source* Author's photographic archive

trends (Ismael 2021). These findings were compared with the empirical outcomes of earlier studies investigating trends in selected settlement and land-use indicators (Karadimitriou et al. 2021). Literature results may corroborate evidence from other European agglomerations (Salvati and Gargiulo Morelli 2014), bringing insights on the debate over the future development of growing (or declining) cities (Keddie and Tonkiss 2010).

Investigating more about the micro-foundations of the phenomenon, households that were risk-loving during the boom have recognized to be more risk-averse during the market downturn (Kaika 2012; Kemeny and Storper 2012; koprowska et al. 2020). In addition, crisis-driven difficulties have made housing less affordable, depressing homeowners' well-being (Kordel and Naumann 2023). Thus, homeownership rates have declined in central cities, and especially among middle- and low-income households (Kourliouros 2003). In a period of shrinking welfare states, and weak national governments, local authorities seem to retake a central role in the metropolitan arena, acquiring new power and the necessary degrees of freedom in managing regional plans, in addition to town master plans (Le Goix 2005; Kroll and Kabisch 2012; Lagarias 2023). A refined understanding of the intimate relationship between local construction markets and housing vulnerability and affordability, is thus increasingly necessary in such contexts, when investigating the consequences of economic stagnation, the associated factors, and the possible coping strategies (Kourliouros 1997; Lee 2011; Kandylis et al. 2012).

5.3 The Mediterranean 'Urban Crisis' as a Form of Economic Recession

Compared to Northern, Central, and Western European counterparts, Southern metropolises underwent urbanization between the late 1940 and the mid-1970s (Leontidou 1996), with compact settlements constituting mono-centric agglomerations organized along metropolitan continuums (Mudu 2006). In several cities, suburbanization started in the late 1970s and alimented a spatially discontinuous and low-density metropolitan expansion (Lindstrom and Bartling 2003; Katsibokis 2013; Liu and Meng 2020). In this region, distinctive socioeconomic, demographic, political, and cultural factors have increasingly shaped development paths for individual cities (e.g. Kazemzadeh-Zow et al. 2017), among which industrial dynamics, social segregation, demographic increase, and a limited participation of citizens and stakeholders to planning decisions, were regarded as pivotal drivers of change, even during the great crisis of the late 2000s (Maloutas 2004; Masini et al. 2019; Luger and Schwarze 2021).

In this perspective, Mediterranean cities have been frequently considered a midpoint between metropolitan agglomerations of affluent and emerging countries, reflected in a progressive transition toward zero (or negative) population growth and aging (Musterd 2005; Orestein et al. 2014; Nickyin et al. 2020). According to Diaz Pacheco and Garcia Palomares (2014), the great economic crisis 'has not braked the metropolitan expansion, but it has affected the model of the city itself', as clearly reflected in a number of statistical indicators analyzed over a sufficiently long time interval. A specific example dealing with long-term population dynamics was illustrated in Fig. 5.3 for metropolitan Athens.

Strategies boosting metropolitan resilience and containing the huge impact of crisis, should revisit recent development modes, re-thinking a sustainable expansion of settlements while rehabilitating vacant land and abandoned (industrial and commercial) spaces more intensively than in the past (Maloutas 2003; Musterd et al. 2017; Mouratidis 2019). Self-contained expansion based on population densification and semi-compact settlements developed on already urbanized areas (assumed as an indirect form of 'brownfield development') were envisaged in previous research as possible—direct or indirect—strategies to contain sprawl in Southern Europe (Maloutas 2007; Salvati et al. 2012; Rubiera-Morollon and Garrido-Yserte 2020). Brownfield development (Fig. 5.4) may be realized, following Diaz Pacheco and Garcia Palomares (2014), with the promotion of land-use diversification at the local scale, and urban design allowing for 'high ratios of sustainable mobility, less based on car commuting and oriented to the public transport'.

In this perspective, Pacione (2003) argued that, especially in the United States, recessions slowed down any process of job decentralization over space. These processes had been occurred mainly in the 2000s, as a consequence of the economic expansion of manufacturing and construction industries (Marmaras 1989). Consequently, the take-up of sustainable transport modes documented to increase, at least moderately, during the last decade (Petsimeris 2002; Pares et al. 2013; Ponstingel

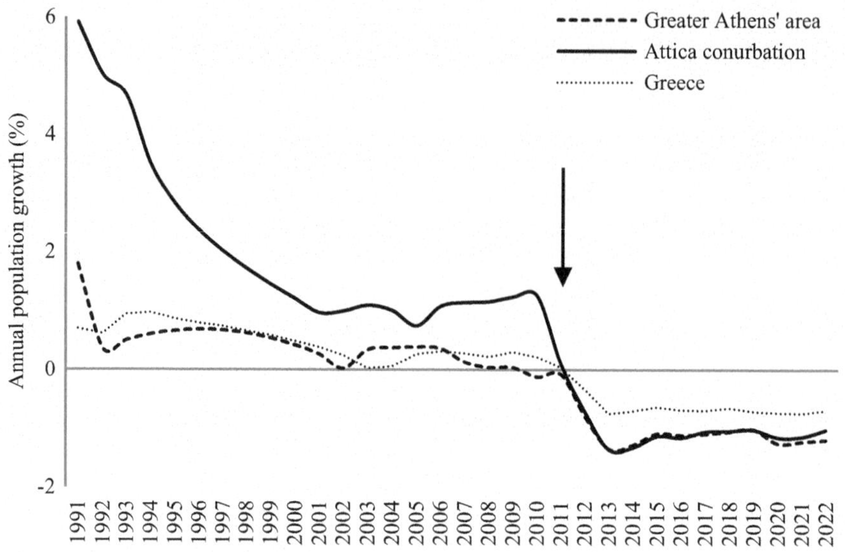

Fig. 5.3 Population dynamics in metropolitan Athens, 1991–2022 (the arrow indicates the beginning of the great recession in Greece). *Source* Author's elaboration of official statistics released from Hellenic Statistical Authority, ELSTAT

2020). In Barcelona, a trend toward a rational use of private modes in favor of shorter trips and non-motorized commuting modes was more recently observed (Munoz 2003; Sauri 2003; Salvati 2013).

5.4 Economic Crisis in the Experience of Mediterranean Metropolises

Distinct spatial patterns in building activity characterized urban and suburban spaces, reflecting divergent responses to recession, and making Mediterranean metropolises as appropriate examples of cities experiencing a 'structural crisis' (Sassen 2010; Kaika 2012; Salvati et al. 2016). These agglomerations are relevant cases when investigating implications and consequences of the great recession on metropolitan growth and settlement sprawl (Salvati 2016; Feleki et al. 2018; Kovacs et al. 2019). Despite the recent expansion of these agglomerations was relatively slow if compared with the increase in urbanization rates observed in both emerging economies and in some dynamic metropolises of wealthier countries (Petsimeris and Tsoulouvis 1997; Power 2001; Richardson and Chang-Hee 2004), the accelerated building cycle observed in the 1990s and the 2000s consolidated dispersed metropolitan structures with spatially

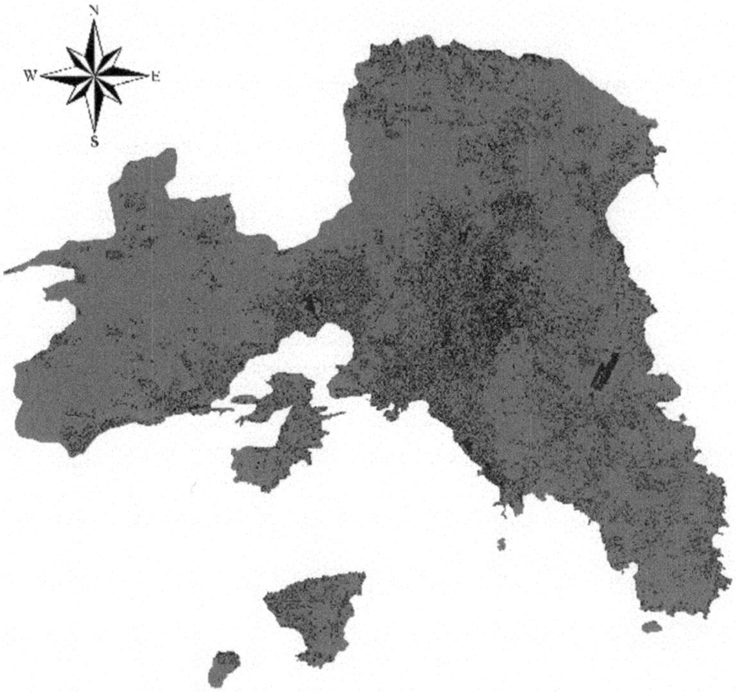

Fig. 5.4 Spatial distribution of brownfield (black) and greenfield (dark gray) settlement development in metropolitan Athens (2006–2018) with the stable land-use matrix illustrated in light gray; source: author's elaboration on Urban Atlas digital maps

discontinuous settlements and low-density villages around some central (and semi-central) economic nodes (Haase et al. 2010; Kabisch and Haase 2011; Gkartzios and Scott 2015).

Decreasing interest rates and increased access to loans fueled a latent 'housing bubble' especially in Spain, Greece and, to some extents, Italy (Lagarias and Prastakos 2020; Mehriar et al. 2020). Infrastructural development and mega-events—such as the Olympic Games in Barcelona and Athens, respectively held in 1992 and 2004, and the 2000 Jubilee in Rome—have presumably boosted such dynamics (Burgel 2004; Remoundou et al. 2016; Roubien 2020). A moderately increasing population, also due to residual migration flows, and the sudden increase in the internal demand for second-homes, contributed to shape such trends (Gavalas et al. 2014; Rontos et al. 2016; Tsilimigkas et al. 2016; Gielen et al. 2018). Pressure for new houses raised by retired population moving from Northern Europe to coastal areas was also locally high, e.g. in Mediterranean Spain (from Andalusia to Catalonia) and moderately relevant also in Greece and Italy, where some flat districts were specifically involved in such dynamics (Schneider et al. 2010; Souliotis 2013; Salvati et al. 2018).

Thanks to a comparative analysis of housing policy and practices (Scott and Storper 2015), additional research contributed to outline a distinctive relationship between the state, markets, civil society, and households in Mediterranean Europe (Hof and Blazquez-Salom 2013; Skayannis 2013; Pagliarin 2018). Homeownership prevailed as a form of social insurance over the life course (Stathakis and Baltas 2024), with public housing often restricted to specific places and interventions with certain population segments (Storper and Scott 2016). Public housing units, especially in Spain and Italy, were frequently sold to the existing tenants (Tomao et al. 2021).

At the same time, the construction sector—mainly in urban areas—underwent a prolonged crisis since 2007, with the cut of the largest residential development projects, and the prevalence of small-scale transformations (or even re-building) downtown (Townsend 2009; Weber 2010; Dewilde and De Decker 2016). New settlements (i.e. greenfield development) were almost exclusively observed as expanding radio-centrically around consolidated economic centers (Fig. 5.5), mimicking the 'compact' dynamics typical of the early 'urbanization' wave (1950s–1960s) in Mediterranean cities (Fekade 2000; Salvati et al. 2013; Filandri and Olagnero 2014). Public investments in the housing sector decreased drastically, and private investments focused on few productive areas—usually along the cities' outskirts (You and Zi 2007), with complete accessibility and medium–low land prices, because of their suboptimal characteristics for both agriculture and residential settlements (Swyngedouw 2004; Chorianopoulos et al. 2014; Salvati and De Rosa 2014).

Fig. 5.5 Suburban landscapes and the compact city in the background. *Source* Author's photographic archive

In Athens, according to empirical evidence from earlier studies (e.g. Maloutas 2007; Chorianopoulos et al. 2010; Kandyli et al. 2012; Alexandri 2015; Panori et al. 2019), long-term building activity has been influenced by multiple factors, including massive immigration, the decline of traditional industries, a weak planning system combined with a 'permissive' building code (Tsiligiannis 2020), and the development of the region's infrastructure that was driven by the 2004 Olympic games (Delladetsima 2006). The dynamic interplay between social classes has also influenced building activity (Malheiros 2002; Arapoglou and Sayas 2009; Gabriel and Rosenthal 2015). Private housing initiatives showed a functional interdependence between multiple property sectors including office, residential, retail, and entertainment (e.g. Salvati et al. 2018), and promoted settlement expansion into a lively—but often economically polarized—suburban environment (Garcia and Riera 2003).

In Barcelona, empirical studies have prefigured future settlement expansion as a result of population growth, densification of economic activities and specific social profiles (Garcia-Ramon and Albet 2000; Salvati et al. 2013; Lagarias and Sayas 2019). As suggested by Nel-lo (2011), population de-concentration has produced suburban spaces with moderately dispersed forms and mixed economic functions (Monclus 2000, 2003; Muniz et al. 2003). Although the highest percentages of buildable land were found in peri-urban municipalities, the most accessible rural areas, especially flat districts along the Mediterranean Sea coast, will experience extensive urbanization in the near future (Marraccini et al. 2015). The highest percentages of buildable land were observed in municipalities with a high frequency of one-person households (Colaninno 2012).

Although regarded as a representative Mediterranean city displaying similar economic and demographic dynamics to other Southern European metropolises (Kaniewski et al. 2013; Salvati and Carlucci 2016; Tsapala et al. 2023), such as Athens and Barcelona, Naples and Istanbul, Lisbon and Seville, Thessaloniki and Marseille, among the others, Rome represents a quite distinct case compared with Spanish and Greek cities as far as pre-crisis and crisis dynamics in the real estate market are concerned (Salvati 2014). Similar to other Italian cities—apart from Milan—Rome, a semi-compact city with discontinuous expansion over two recent decades (from the mid-1980s to the mid-2000s), was relatively less exposed to the 'housing bubble' compared with many Spanish counterparts (Salvati et al. 2019). Conversely, the eternal city has still been experiencing a moderate (and compact) exurban development with a latent wave of low-density residential settlement expansion at the same time (Garcia 2010; Sharaf et al. 2018; Coq-Huelva and Asian-Chaves 2019). Even if the overall impact of crisis was probably lower in Italy than in Greece, recession had important effects, e.g. on local job markets, with unemployment rates rising by 7 percent points over five years (from 7% in 2007—the apex of the economic expansion starting in the late 1990s—to 14% in 2012). Population and economic transformations observed in Rome before, during, and after the 2007–2010 crisis may thus shed further light in the general urbanization pattern observed at the Mediterranean scale (e.g. Mudu 2006; De Muro et al. 2011; Salvati 2014).

While arguing the existence of hybrid 'metropolitan models' (e.g. placed in-between hyper-sprawled and compact settlements), Diaz Pacheco and Garcia Palomares (2014) contributed operationally to the assessment of morphological changes in patterns and processes of urban growth in relation with the great recession in Madrid, considering the whole metropolitan region. Based on their work, similarities and differences in short-term population growth, urban expansion modes, and settlement attributes before and during recession, revealed to be meaningful information shedding light on the most recent transformations in the Mediterranean region (e.g. Recanatesi et al. 2016; Smiraglia et al. 2016; Vinci et al. 2023). Despite their deep research effort, however, evidence on the recession's impact on metropolitan growth characteristics and spatial directions remained quite occasional, and basically restricted to some local contexts (Vaiou 2002; Capello et al. 2015; Zambon et al. 2018a, b). More recently, the real estate sector has dealt with unexpected problems and uncertain prospects, especially under the Covid-19 pandemics (Vicino et al. 2022), bringing new planning issues and housing strategies to the fore (Vaiou 1997; Vradis 2014; Tsilimingkas et al. 2022).

5.5 Fringe Dynamics and the Economic Crisis

Peri-urban landscapes are among those landscapes identified as 'ordinary' (Fig. 5.6), or apparently 'without quality', or with a latent quality that is not easily and immediately recognized according with traditional aesthetic, perceptive, or ecological criteria (van Criekingen 2010; Vidal et al. 2011; Tsilimingkas and Kizos 2014). The widespread idea of a 'bucolic' landscape, since the ancient Greek and Roman ages, influences the overall (negative) judgment inherent in peri-urban environments (Lagarias and Sayas 2018; Zambon et al. 2018a, 2018b, 2019). This makes the historic, cultural, social identity potentials of these landscapes almost negligible, depressing in turn their agricultural potential (Degen and Garcia 2012; Masoumi 2014; Salem et al. 2020). Suburban landscapes are intended as fragmented and complex mosaics with a mixed and spatially varying land-use (Mayer and Somerville 2000; Nelson et al. 2004; Sampson et al. 2017). Agriculture alternates with residential urban fabric, industrial complexes and districts, derelict/abandoned/vacant land, in a sometimes incoherent territorial structure leading to a sort of 'landscape mix' (Johnson 2001; Doygun 2009; Soulard et al. 2018).

However, fringes are still strategic places contributing to ecosystem equilibrium, controlling metropolitan expansion and establishing original city-countryside relationships (Herold et al. 2002, 2003, 2005; Salvati and Zitti 2009). At the same time, fringes are territories where urban and rural lifestyles contaminate and merge, having an increased population density and declining job rates in the primary sector as distinctive characteristics in respect with the surrounding urban and rural districts (Henderson 1991; Garcia-Ayllon 2013; Gil-Alonso et al. 2016). After decades of indiscriminate exploitation of soil resources in the old continent, suburban territories

Fig. 5.6 Economic and landscape centrality of commercial harbors in Naples Riviera, reflecting what can be considered a sort of 'ordinary' landscape for Mediterranean cities. *Source* Author's photographic archive

appear to be fragile environments where land degradation processes are more pervasive and widespread than elsewhere in the same bio-geographical region (Polyzos et al. 2013; Martì et al. 2017). Territorial contexts and socioeconomic determinants form the basis of landscape degradation, taken as a result of anthropogenic transformations leading to potentially serious 'desertification' processes, being particularly common in the Mediterranean region (Petrov et al. 2009; Palang and Peil 2010).

Is fringe including mostly deteriorated landscapes? A careful observation may reveal, in such a perspective, the countless signs that a rapid economic development has left in Mediterranean landscapes (Marshall 1995; Parcerisas et al. 2012; Pace et al. 2023). Since World War II, Mediterranean fringe land has left the connotation of the widespread 'rurality' typical up to the first half of the last century, becoming progressively urbanized and industrialized (Muniz and Garcia-Lopez 2010; Escudero Gomez et al. 2019; Schwaab 2022). A seamless urban fabric has invaded coastal areas. Intensive cropping systems inland and livestock industry became increasingly mixed with high-density service settlements, manipulating the typical features of rural landscapes, and reducing together aesthetic and ecological qualities—some of the most promising capitals for future growth (McMillen and McDonald 1993). The 'territorial component' of current socioeconomic scenarios became more important in any economic analysis, assuming that investments go from being subsidiary to a trusting spirit in local, environmental, and entrepreneurial system resources (Phillips and Goodstein 2000; Nilsson et al. 2013).

Different evaluation methods have been developed integrating several target variables via quantitative approaches, and produced summary indexes and other computational approaches assessing the level of land degradation (Serra et al. 2008; Sanz et al. 2013; Amato et al. 2015). Land vulnerability in the Mediterranean region was basically estimated considering together climate, soil, and vegetation qualities, in addition to anthropogenic factors affecting population density and growth, as well as the level of land management and environmental protection of a given territory (Silva and Clarke 2002; Solecki and Oliveri 2004; Song and Knaap 2004). Such evaluations are of particular interest when referring to a diachronic evolution of ecological processes, or to short-term projections of future environmental conditions (Villa 2012; Christodoulou and Oikonomou 2018). Degradation syndromes can thus be evaluated for the past, and scenarios developed for the near future, providing a basis for elaboration of sustainable land management strategies (Veldkamp and Verburg 2004; Stewart 2008; Vaz et al. 2015).

At the same time, a qualitative assessment of land use, together with a visual analysis of landscapes, seems to provide essential information interpreting complex and multidimensional changes that have taken place in the past. These processes substantially contribute to understand contemporary dynamics, reinforcing the confidence for future projections (Wynn 1984; Uotila et al. 2005; Wiek et al. 2006; van Vliet et al. 2013; Triantakonstantis and Stathakis 2015). Reflecting in turn a persistent conflict between social and economic dimensions, land degradation is structurally related to production and consumption systems based on a long-term expansion of marketable goods (Dawkins and Nelson 2002; Xiang and Clarke 2003; Koomen 2007). These dynamics assume the principle that economic growth would automatically generate an increase in social welfare (Shearer 2005; Leontidou 2009; Habibi and Asadi 2011; Varela-Cardamio et al. 2019). Changes in the direction of a different model of progress and development based on sustainability principles are clearly necessary and not delegable to future generations.

5.6 'Crisis' Landscapes and (Sub)urban Sustainability

Sustainable conditions in the economic, social, and environmental dimensions of regional growth, require the maintenance of the actual stock of natural and human capitals, representing a substantial challenge in sociocultural assets and management perspectives (Silva 2004; Smith 2005; Wey and Hsu 2014). In a progressive transition to a low-carbon economy, besides the technical and policy choices, the spatial actors involved in decision-making processes play a pivotal role via transformation of their modes of production, consumption, and lifestyles (Kloosterman and Musterd 2001). They can promote a cultural break-point overcoming unsustainable patterns of land resources' use, and reducing the environmental impact of human activities, so that the ethic of 'sustainable development' become culture, political action and, together, economic practice (Westerink et al. 2013; Tsiligkas et al. 2018; Rejeb Bouzgarrou et al. 2019).

In this perspective, landscape degradation is one of the most important environmental issues that our societies are called upon to face (Slaughter 2002), because of the serious consequences it poses for human health and the environment. On the one hand, this issue has often been at the core of media focus, stimulating debates in decision-makers and the public opinion at large (Sterk et al. 2011). On the other hand, the volatile interest on such issues, being sometimes associated (erroneously) with natural disasters and extreme events, prevents the consolidation of environmental awareness in local communities (Paul and Tonts 2005). These events have therefore channeled the attention of the general public toward 'hegemonic' (and easily to communicate) concepts such as the desertification-climate relationship, posing attention to the biophysical factors at the base of climate change, e.g. local warming, drought, water scarcity, and soil aridity—while neglecting the important role of social, economic, cultural, political, and institutional factors (Saizen et al. 2006; Overmars et al. 2007; Munoz 2010).

Landscape degradation, however, has been put to the forefront of recent environmental dynamics at various observation scales (Fig. 5.7), requiring scientifically sound approaches and a more conscious and less sensationalist dissemination plan (Verburg et al. 2004; Weng 2007; Sakieh et al. 2015). Moreover, the issue would require continuous and systematic awareness through permanent monitoring of life quality and ecosystem stability, and the subsequent development of short- and long-term (calibrated) intervention strategies (Ratcliffe and Krawczyk 2011). A framework of this magnitude is made even more complicated by the drastic reduction, for public budget constraints, of public investment driven by economic stagnations (Polet and Monden 1997). The psychological effects of a growth sustained from above—the protagonist of past territorial development—have decreased, impacting marginal landscapes preserved from land speculation and soil sealing (Puglisi and Marvin 2002; Hennig et al. 2015; e Silva and Correia 2023).

Rural quality and biodiversity features were compromised both during economic booms and stagnations, displaying a spatially widespread pattern of resource consumption that shapes rural landscapes alternating high-intensity areas with large farms at the limit of abandonment (Rounsevell et al. 2005; Rienow and Goetzke 2015). The possibility of re-balancing unsustainable fringe districts is basically associated with the capacity to adopt eco-compatible production systems for intense cropping and the possibility of making farm production in marginal areas viable in the medium-term (Basnou et al. 2013; Newman et al. 2014; van Vliet et al. 2016).

Earlier studies on sustainable agricultural systems have evidenced the importance of considering together three dimensions (economic, social, and environmental) adopting a 'truly systemic' point of view (e.g. Al-Kheder et al. 2008). In the first dimension, the efficient use of resources and the vitality of the agricultural sector should be identified as key research and policy issues, together with the contribution of the primary sector to the operational conservation of rural areas (Sharma et al. 2006). The social dimension in turn refers to human capital and its differential features at the local scale (Sardà et al. 2005). Finally, the environmental dimension highlights the appropriateness of natural resources' management solutions stabilizing landscape, water, and soil quality, demonstrating that integrated and systematic evaluations provide correct and complete interpretations of landscape trajectories (Serra et al. 2014; Duvernoy et al. 2018; Perrin et al. 2018).

Governing landscape changes for the speed and intensity of ongoing changes determines macroscopic and severe effects whose consequences are evident with all the implications of related policies (Turok and Mykhnenko 2007). Climate change

Fig. 5.7 Coexistence of residual agriculture (subsidence vineyards) in a context of degraded soils, with recently installed solar plants in Grammatiko, Northern Attica. *Source* Author's photographic archive

was demonstrated to involve social dynamics and economic productivity. Considering landscape matrices as complex systems along peri-urban fringes, one can perceive how the notion of 'change', 'variability', 'anomaly', and 'balance' should relate to collective (territorial) dimensions and economic spheres, implying the global involvement of multiple variables and factors of change (Schneider and Woodcock 2008). It is essential to build a knowledge-based (preventive) approach that enables decision-makers to develop the most effective environmental policies and to feed public debates and scientific knowledge (Samet 2013). A permanent monitoring becomes fundamental to effectively manage territorial resources and mitigate the effects of activities leading to land degradation (Barbero-Sierra et al. 2013; Serra et al. 2013; Wolff et al. 2020). With such feedbacks, the demand of operational tools assessing the latent relationship between local territories is particularly evident (Fig. 5.8), being based on multiple expressions of transforming spaces, complex social systems, and environmental dynamics (Salvati 2013; Rijkens-Klomp and Van Der Duin 2014; Marraccini et al. 2015).

In a context where economy, society, and the environment became more and more integrated and interconnected, the issue of sustainable development becomes increasingly topical at the local level (Tsai 2005; Burger and Meijers 2012). The debate on the organization and infrastructure of territories is strictly related with

Fig. 5.8 Enlightened waterfronts and the darkness of crisis in coastal Scario, Southern Italy; author's photographic archive

such concepts, having operational relevance in socioeconomic contexts that featured sprawl and landscape fragmentation as their main morphological traits (Mills 1967; Rappert 1999; Longhi and Musolesi 2007). On the one hand, it was demonstrated how development engines have been based, in the immediate aftermath of World War II, on the consolidation of manufacturing industry, sometimes concentrated in the most accessible districts, and sometimes fragmented in barely unspecialized districts (Saizen et al. 2006). On the other hand, manufacturing districts have more recently exerted a significant impact on local competitiveness (Salvati and Carlucci 2014). While the network of intermediate towns has come to expand significantly in the near past—producing relevant shares of country's wealth—nowadays two are the major directions of local development: (i) innovation functions capable of engaging new, and often delocalized, globalization districts, and (ii) a new territorial dimension of 'socioenvironmental' quality, associated with municipalities (and districts) with relevant cultural and ecological endowments deriving from a diffused, civic sense of belonging that indirectly boosts some specialized domains in the production of niche goods, namely in the primary sector (MacKinnon et al. 2009). In this dimension, district specialization can also emerge in the new wellness industry, assuring high-quality welfare services, and high-quality goods with a great food tradition. Rural hospitality and environmental sustainability will be thus enhanced as opposed to the paradigm of population concentration, scale/agglomeration economies, and increased accessibility (Camagni et al. 2002).

5.7 The Inherent Crisis in Contemporary 'Metropolitan Models'

To overcome crisis, the challenge of peri-urban fringe means safeguarding the environment and making agriculture viable, promoting a new development model based on economic activities such as cultural tourism and agro-tourism (Cabral et al. 2011). The integrity and authenticity of fringe landscapes (Fig. 5.9) became a development engine based on their multifunctional values that meet the demand for new services (Aguilar 2008). Local initiatives and policies should subsidize and promote.

(i) eco-friendly markets with collective buyers,
(ii) teaching activities (didactic farms, agro-kindergartens, natural parks, trekking),
(iii) accommodation facilities and farm equipment supporting recreational activities,
(iv) social structures (e.g. social farms, community for sociosanitary recovery, hobby farming), and
(v) direct sales facilities of local agricultural (food and non-food) products.

Close proximity to cities allows agricultural entrepreneurs to develop a competitive advantage for the spread of land products on local markets, providing zero-kilometer food, with significant transport savings and ecosystem benefits (Alberti 1999; Coiacetto 2006; Diappi 2013).

A 'socially responsible territory' can be finally promoted with appropriately oriented and tuned finely participatory paths, contributing to local communities fully aware of the socioeconomic value of natural resources and actively involved in giving value to local peculiarities through building cooperative and cohesive actions linking them with local government (Anthony et al. 2004; Almeida et al. 2013; Vinci et al.

Fig. 5.9 Mediterranean horizons: natural habitats, rural spaces, and the sea coast in Policastro, Southern Italy. *Source* Authors' own photographic archive

2022). This virtuous development path coincides with an enhanced perception of place identity, with the aim at improving quality of life within a community-based planning vision (Abelairas-Etxebarria and Astorkiza 2012). 'Territorial foresight' actions allow the involvement of all actors with the aim at harmonizing the different stakeholders' views, creating a network that links local stakeholders, communities and contexts, to promote cultures and practices leading to local cooperation (Sicular et al. 2007). In this way, more sustainable societies with enterprise models and solutions that address local needs using knowledge and resources may share common development prospects (Chaudhuri and Clarke 2012). Convergence on landscape conservation targets can lead to processes of social innovation and trigger the economic development of peri-urban territories (Gantar and Golobic 2015; Zwiers et al. 2016; Zambon et al. 2017).

At the same time, moving from cities to rural areas within large metropolitan regions in Southern Europe, territorial processes and local products promote social cohesion, a 'sense of belonging' to the place in local communities, tourism development, and an effective territorial marketing (Casellas 2006). 'Landscape strategies' in peri-urban planning projects recognize the strong identity of places and values better attained to local communities, giving value to the related products (Dammers 2010). Local governance actions should consider the different needs of the involved actors, creating a network that may link stakeholders and local contexts, highlighting the possible alternatives to intensive (and irreversible) exploitation paths (Dorning et al. 2015). Building on the sustainable development of cities, there is a need for balancing anthropogenic pressure with the use of resources (Fig. 5.10), definitely ensuring the balance of human needs with natural capital endowments (Dantzig 1999).

5.8 Discussion

Sociodemographic change paralleled economic processes that influence settlement configuration, sociospatial gradients, and the distribution of productive functions over metropolitan areas. With this perspective in mind, it is important to outline that sustainable development promotes the safeguard of land resources, considering in turn the principles of solidarity and cohesion to improve the future quality of the environment within the economic dynamics of production and consumption (Dufva et al. 2015), in cities as well as in the surrounding metropolitan regions. Interventions designed to make landscape transformations as 'sustainable' can be applied and yield satisfactory results in mitigating land degradation in peri-urban areas (Zambon and Salvati 2019) when the territorial actors perceive such interventions as a support to monetary profitability (Hall 1997). Moreover, these measures are timely in those areas where agricultural production represents a significant contribution to the production of local wealth. Regulating and protecting rural areas in peri-urban districts requires a careful design of integrated policies that balance constraints and incentives, mainly built through the involvement and participation of farmers,

Fig. 5.10 Degraded landscapes, climate aridity, and soil erosion in Malaga fringe; authors' own photographic archive

entrepreneurs, and citizens. A particularly important task is to define and give incentive to multifunctional agricultural enterprises along metropolitan fringes, with the aim at supplying goods and environmental services to local communities (Haslauer 2012). In this way, agricultural entrepreneurs are recognized with an important function of safeguarding rural landscapes, maintaining hydro-geological functioning and promoting the productive vocation of territories (e.g. Fox et al. 2012).

Areas in-between towns and the countryside may finally provide a valid alternative to metropolitan development offering leisure services, without having to face the high cost of construction and maintenance of commons (Bourne 1996; Gordon and Richardson 1996; Cakir et al. 2008). In addition, interstitial and peri-urban rural areas have linked the cultural values of a 'wild rurality' with the historical identity of natural landscapes and traditional crops (Delfanti et al. 2016). These approaches place local territories in a perspective of sustainable development because they intrinsically recognize, plan, and protect identity values and cultures of local communities (Newman and Kenworthy 1999), promoting initiatives that enhance the peculiarities of places, involving citizens in active conservation strategies (Briassoulis 2005).

The picture is complicated by rapid changes in societies and transformations in the economic structure, exerting huge impacts on the spatial organization of entire regions. Just to look at some recent trends, the 'housing bubble' recorded in Southern

Europe during the early-2000s and the mid-2000s, and the subsequent financial crisis that struck violently the same countries, both determined uneven settlement expansion because of land speculation, second home, and tourism development, a residual industrial growth, and massive infrastructural improvements, with drastic consequences in the use of rural land surrounding cities. At the same time, the continuous decline of traditional industries led to original models of urban expansion leading to largely decentralized spatial organizations of cities and regions, while privileging local-scale development in some cases.

The partial gap in the current body of knowledge that specifically deals with the intrinsic relationship between metropolitan change and the local transformations in the socioeconomic base of cities, is somewhat surprising given that urbanization still prevails as an effective factor shaping structure and functions of regions. An advanced discussion of urban patterns—either in the form of compact growth and in the way sprawl is taking ground in some peripheral countries of the old continent—and the consequent transformations in the production base of regions, is also timely, given that Mediterranean Europe was experiencing a convulse transition from mono-centric cities to a moderately polycentric system of large and medium economic poles, influencing districts further away from the traditional urban centers.

Socioeconomic dynamics in the academic literature have been routinely investigated using indicators, quali-quantitative analysis' approaches, narratives, and visual perspectives varying largely in time and space. Overcoming a relatively balanced population distribution over a millenary history, European regions have featured a long-lasting expansion alimented by positive rates of natural growth and massive immigration. Widening the socioeconomic divide in urban and rural areas, such dynamics have consolidated the decline of peripheral areas, exemplified in the geographical gradient from coastal to inland districts. A demographic re-distribution over broader regions, likely intended the most frequent outcome of this kind of study, was usually the result of intrinsic alterations in the density gradient between urban and rural locations, being in turn influenced by local dynamics associated with sequential expansion and stagnation waves.

Whereas a number of European regions still experience a continuous process of urban growth, a non-negligible set of local contexts underwent economic decline (and/or demographic shrinkage) over recent decades, and some other regions have displayed less pronounced trajectories, positioning in-between these two poles of expansion or recession. In this perspective, spatial demography is a fundamental factor contributing to metropolitan shrinkage. Official statistics documented how several European cities at the beginning of the twenty-first century have experienced a population decrease, mainly downtown and, less frequently, in ring districts. More recently, some of the most important Mediterranean cities seem to be not completely out of this trend, after many decades of continuous increase of resident population and densities.

Socioeconomic changes in those cities are receiving increased attention in both academic and political debates, since they imply a silent transformation of traditional residential patterns and neighborhoods' profile. At the same time, recent empirical evidence from narrative and visual approaches started identifying early signals of urban re-polarization and sociodemographic diversification that are still relatively latent in the latest data elaborated from population registers, census records, and other official statistics. Based on the inherent diversification in several (underlying) factors, the results of empirical studies provide mixed and heterogeneous evidence on the effect of the great recession on metropolitan dynamics in Southern Europe.

While cyclical urbanization models seem to be no longer able to capture and interpret the even more complex dynamics of metropolitan development in the old continent, it is progressively clearer how cities and agglomerations do not follow individual evolutionary stages of growth. As a matter of fact, vastly different development waves coexist in the same city and metropolitan region. The importance of regions in terms of economic growth and employment, and their attractiveness to internal and international migrants, is also demonstrated in the recent history of Mediterranean cities. Although continuing migration to economically dynamic districts is an evergreen challenge for spatial planning, sociodemographic trends toward declining household size and the splitting up of households, are recently re-orienting the physical growth of settlements toward possibly unexpected paths, altering the macroeconomic demand for housing, services, and infrastructure.

Based on this assumption, we positioned here at the interface between spatial demography, population geography, urban studies, and regional science, with the aim at discussing territorial gaps as a function of the local socioeconomic context at different stages of the economic cycle (basically, expansion and recession). Assessing the influence of economic expansion and recession on building activity has pivotal implications for the design of planning measures and effective local development policies. Empirical studies were increasingly required to identify compact and sprawled urbanization as a result of changes in household preferences, public policies, and the role of private agents during economic booms and stagnations.

5.9 Concluding Remarks

While approaching the conclusion of our interpretative path in the recent history and geography of Mediterranean cities, we cannot demise how relevant socioeconomic gaps in Southern Europe have been still found among neighboring regions, and a part of marginal, inland areas is still associated with stagnant economic systems. Metropolitan growth as well as shrinkage processes are typically related to specific contexts—cut across national, regional, and local differences in culture, governmental structure, and policy—being installed in distinctive interpretative frameworks, based around diverging normative settings, and ultimately leading to different policy implications. Research focusing on macroscale factors (metropolitan hierarchies, economic structures, social dynamics, market-state interactions, and institutional change) has clearly indicated how territorial disparities persisted over the last decades, and are apparently more intense in regions with a traditional economic base, population aging, low-quality human capital, and emigration. Urban sprawl is evidently one of the most powerful drivers of socioeconomic inequalities.

Use of homogeneous sets of structural and functional indicators—monitoring changes over time in home-building activity and the underlying socioeconomic context—is increasingly required in this perspective, especially when ascertain the intrinsic nexus between urbanization and the formation (or consolidation) of regional disparities. It is well-known how, for instance, cities and regions facing adverse sociodemographic trends and stagnant economic dynamics appear to be increasingly vulnerable to increasing fiscal constraints, thereby limiting their ability to effectively and pragmatically manage urban decline.

The reason for such research advancements in regional science—as they are stemming from this book, as a sort of take-home message for any scholar interested in urban theory and practice—is inherent in the fact that an improved design of policies addressing economic growth under local sustainability and metropolitan resilience principles may benefit from a better understanding of spatial direction and intensity of recessions, and development patterns and trends. Sustainability and resilience are basic elements of any development path oriented toward social cohesion and environmental neutrality. In these regards, understanding how sociodemographic processes, building cycles, and economic downturns, interact together—shaping regional competitiveness, location attractiveness, metropolitan sustainability, and resilience to external shocks—contributes to a better identification of factors promoting local development and environmental neutrality.

References

Abelairas-Etxebarria P, Astorkiza I (2012) Are land use policies preserving farmland from urban sprawl? Rev Econ Stud 4(5):24–29

Aguilar AG (2008) Peri-urbanization, illegal settlements and environmental impact in Mexico City. Cities 25:133–145

Aguilera F, Valenzuela LM, Botequilha-Leitão A (2011) Landscape metrics in the analysis of urban land use patterns: a case study in a Spanish metropolitan area. Landsc Urban Plan 99:226–238

Alberti M (1999) Modeling the urban ecosystem: a conceptual framework. Environ Plann B Plann des 26:605–630

Alexandri G (2015) Reading between the lines: gentrification tendencies and issues of urban fear in the midst of Athens' crisis. Urban Stud 52(9):1631–1646

Al-Kheder S, Wang J, Shan J (2008) Fuzzy inference guided cellular automata urban-growth modelling using multi-temporal satellite images. Int J Geogr Inf Sci 22(11):1271–1293

Allen J, Barlow J, Maloutas T, Padovan L (2004) Housing in southern Europe. Blackwell, London

Almeida J, Condessa B, Pinto P, Ferreira JA (2013) Municipal urbanization tax and land-use management—the case of Tomar, Portugal. Land Use Policy 31:336–346

Amato F, Pontrandolfi P, Murgante B (2015) Supporting planning activities with the assessment and the prediction of urban sprawl using spatio-temporal analysis. Ecol. Inf. 30:365–378

Andreotti A, Le Galès P, Moreno Fuentes FJ (2013) Controlling the urban fabric: the complex game of distance and proximity in European upper-middle-class residential strategies. Int J Urban Reg Res 37(2):576–597

Angel S, Parent J, Civco DL, Blei A, Potere D (2011) The dimensions of global urban expansion: Estimates and projections for all countries, 2000–2050. Prog Plan 75(2):53–107

Anthony J (2004) Do state growth management regulations reduce sprawl? Urban Affairs Rev 39(3):376–397

Arapoglou VP, Sayas J (2009) New facets of urban segregation in Southern Europe. Eur Urban Reg Stud 16(4):345–362
Arbaci S (2008) (Re)Viewing ethnic residential segregation in southern European cities: housing and urban regimes as mechanisms of marginalisation. Hous Stud 23(4):589–613
Arbaci S, Malheiros J (2010) De-segregation, peripheralisation and the social exclusion of immigrants: southern European cities in the 1990s. J Ethn Migr Stud 36(2):227–255
Arribas-Bel D, Nijkamp P, Scholten H (2011) Multidimensional urban sprawl in Europe: a self-organizing map approach. Comput Environ Urban Syst 35(4):263–275
Bajocco S, De Angelis A, Salvati L (2012) A satellite-based green index as a proxy for vegetation cover quality in a Mediterranean region. Ecol Ind 23:578–587
Barbero-Sierra C, Marques MJ, Ruíz-Pérez M (2013) The case of urban sprawl in Spain as an active and irreversible driving force for desertification. J Arid Environ 90:95–102
Barras R (2009) Building cycles: growth and instability. Wiley, London
Basnou C, Álvarez E, Bagaria G, Guardiola M, Isern R, Vicente P, Pino J (2013) Spatial patterns of land use changes across a Mediterranean metropolitan landscape: implications for biodiversity management. Environ Manage 52:971–980
Batty M, Xie Y (1994) From cells to cities. Environ Plann B Plann des 21:31–48
Biber E, O'Neill M (2021) Building to burn? Permitting exurban housing development in high fire hazard zones. Ecol LQ 48:943
Bourne L (1996) Reinventing the suburbs: old myths and new realities. Prog Plan 4(3):163–184
Brade I, Herfert G, Wiest K (2009) Recent trends and future prospects of socio-spatial differentiation in urban regions of Central and Eastern Europe: a lull before the storm? Cities 26:233–244
Brenner N, Marcuse P, Mayer M (2012) Cities for people, not for profit: critical urban theory and the right to the city. Routledge, London
Briassoulis H (2005) Policy integration for complex environmental problems. Ashgate, Aldershot
Bruegmann R (2019) Sprawl: a compact history. University of Chicago Press, Chicago
Bueno-Suárez C, Coq-Huelva D (2020) Sustaining what is unsustainable: a review of urban sprawl and urban socio-environmental policies in North America and Western
Burgel G (2004) Athènes, de la balkanisation à la mondialisation. Méditerranée 103(3–4):59–63
Burger M, Meijers E (2012) Form follows function? Linking morphological and functional polycentricity. Urban Studies 49(5):1127–1149
Cabral P, Santos JA, Augusto G (2011) Monitoring urban sprawl and the national ecological reserve in Sintra-Cascais, Portugal: Multiple OLS linear regression model evaluation. J Urban Plann Dev 137:346–353
Cakir G, Un C, Baskent EZ, Kose S, Sivrikaya F, Keles S (2008) Evaluating urbanization, fragmentation and land use/cover change pattern in Istanbul city, Turkey from 1971 to 2002. Land Degrad Dev 19:663–675
Camagni R, Gibelli MC, Rigamonti P (2002) Urban mobility and urban form: the social and environmental costs of different patterns of urban expansion. Ecol Econ 40:199–216
Capello R, Caragliu A, Fratesi U (2015) Spatial heterogeneity in the costs of the economic crisis in Europe: are cities sources of regional resilience? J Econ Geogr 15(5):951–972
Carlucci M, Grigoriadis E, Rontos K, Salvati L (2017) Revisiting a hegemonic concept: long-term 'Mediterranean urbanization' in between city re-polarization and metropolitan decline. Appl Spat Anal Policy 10:347–362
Carlucci M, Chelli FM, Salvati L (2018) Toward a new cycle: Short-term population dynamics, gentrification, and re-urbanization of Milan (Italy). Sustainability 10(9):3014
Casellas A (2006) Las limitaciones del 'modelo Barcelona': una lectura desde el Urban Regime Analysis. Documents D'anàlisi Geografica 48:61–81
Cassiers T, Kesteloot C (2012) Socio-spatial inequalities and social cohesion in European cities. Urban Stud 49(9):1909–1924
Catalán B, Saurí D, Serra P (2008) Urban sprawl in the Mediterranean? Patterns of growth and change in the Barcelona Metropolitan Region 1993–2000. Landsc Urban Plan 85(3–4):174–184

Cazcarro I, Duarte R, Sánchez-Chóliz J (2012) Economic growth and the evolution of water consumption in Spain: a structural decomposition analysis. Ecol Econ 96:51–61
Cebrián Abellán F, Sánchez Ondoño I (2019) Urban sprawl in inner medium-sized cities: the behaviour in some Spanish cases since the beginning of the 21st century. Urban Sci 3(1):10
Chaudhuri G, Clarke KC (2012) How does land use policy affect urban growth? A case study on the Italo-Slovenian border. J Land Use Sci 7
Cho S-H, Kim SG, Roberts RK, Lambert DM, Kim T (2014) Effects of land-related policies on land development during a real estate boom and a recession. Growth Chang. https://doi.org/10.1111/grow.12071
Chorianopoulos I, Pagonis T, Koukoulas S, Drymoniti S (2010) Planning, competitiveness and sprawl in the Mediterranean city: the case of Athens. Cities 27:249–259
Chorianopoulos I, Tsilimigkas G, Koukoulas S, Balatsos T (2014) The shift to competitiveness and a new phase of sprawl in the Mediterranean city: enterprises guiding growth in Messoghia-Athens. Cities 39:133–143
Christodoulou C, Oikonomou M (2018) Urban sprawl typologies in medium-sized Greek cities. A qualitative morphological analysis in the peri-urban landscape. Urban Morphol 16:18
Ciommi M, Chelli FM, Carlucci M, Salvati L (2018) Urban growth and demographic dynamics in southern Europe: toward a new statistical approach to regional science. Sustainability 10(8):2765
Ciommi M, Chelli FM, Salvati L (2019) Integrating parametric and non-parametric multivariate analysis of urban growth and commuting patterns in a European metropolitan area. Qual Quant 53(2):957–979
Coccossis H, Economou D, Petrakos G (2005) The ESDP relevance to a distant partner: Greece. Eur Plan Stud 13(2):253–264
Cohen B (2006) Urbanization in developing countries: current trends, future projections, and key challenges for sustainability. Technol Soc 28:63–80
Coiacetto E (2006) Real estate development industry structure: consequences for urban planning and development. Plan Pract Res 21(4):423–441
Colaninno N (2012). Sustainable territorial development and urban growth: a critical interaction: the Spanish Mediterranean Coast, and Catalonia, during the last two decadas. ACE Arch City Environ 7(20):149–172
Coq-Huelva D, Asián-Chaves R (2019) Urban sprawl and sustainable urban policies. A review of the cases of Lima, Mexico City and Santiago de Chile. Sustainability 11(20):5835
Couch C, Petschel-Held G, Leontidou L (2007) Urban sprawl in Europe: landscapes, land-use change and policy. Blackwell, London
Cuadrado-Ciuraneta S, Durà-Guimerà A, Salvati L (2017) Not only tourism: unravelling suburbanization, second-home expansion and "rural" sprawl in Catalonia, Spain. Urban Geogr 38(1):66–89
Dammers E (2010) Making territorial scenarios for Europe. Futures 42:785–793
Dantzig GB (1999) Planning under uncertainty. Ann Oper Res 85(0)
Darques R (2015) Mediterranean cities under fire. A critical approach to the wildland–urban interface. Appl Geogr 59:10–21
Dawkins CJ, Nelson AC (2002) Urban containment policies and housing prices: an international comparison with implications for future research. Land Use Policy 19:1–12
De Rosa S, Salvati L (2016) Beyond a 'side street story'? Naples from spontaneous centrality to entropic polycentricism, towards a 'crisis city.' Cities 51:74–83
De Muro P, Monni S, Tridico P (2011) Knowledge-based economy and social exclusion: shadow and light in the Roman socio-economic model. Int J Urban Reg Res 35(6):1212–1238
Degen M, García M (2012) The transformation of the 'Barcelona model': an analysis of culture, urban regeneration and governance. Int J Urban Reg Res 36(5):1022–1038
Delfanti L, Colantoni A, Recanatesi F, Bencardino M, Sateriano A, Zambon I, Salvati L (2016) Solar plants, environmental degradation and local socioeconomic contexts: a case study in a Mediterranean country. Environ Impact Assess Rev 61:88–93

Delladetsima PM (2006) The emerging property development pattern in Greece and its impact on spatial development. Eur Urban Reg Stud 13(3):245–278
Dewilde C, De Decker P (2016) Changing inequalities in housing outcomes across western Europe. Hous Theory Soc 33(2):121–161
Di Feliciantonio C, Salvati L (2015) 'Southern' alternatives of urban diffusion: investigating settlement characteristics and socio-economic patterns in three Mediterranean regions. Tijdschr Econ Soc Geogr 106(4):453–470
Di Feliciantonio C, Salvati L, Sarantakou E, Rontos K (2018) Class diversification, economic growth and urban sprawl: evidences from a pre-crisis European city. Qual Quant 52(4):1501–1522
Diappi L (2013) Emergent phenomena in housing markets. Gentrification, housing search, polarization. Springer, Berlin
Díaz-Pacheco J, García-Palomares JC (2014) Urban sprawl in the Mediterranean urban regions in Europe and the crisis effect on the urban land development: Madrid as study case. Urban Stud Res
Domene E, Sauri D (2006) Urbanization and water consumption: influential factors in the metropolitan region of Barcelona. Urban Stud 43(9):1605–1623
Domene E, Sauri D, Parés M (2005) Urbanization and sustainable resource use: the case of garden watering in the metropolitan region of Barcelona. Urban Geogr 26(6):520–535
Dorning MA, Koch J, Shoemaker DA, Meentemeyer RK (2015) Simulating urbanization scenarios reveals tradeoffs between conservation planning strategies. Landsc Urban Plan 136:28–39
Doygun H (2009) Effects of urban sprawl on agricultural land: a case study of Kahramanmaraş, Turkey. Environ Monit Assess 158:471–478
Dufva M, Könnölä T, Koivisto R (2015) Multi-layered foresight: Lessons from regional foresight in Chile. Futures 73:100–111
Durà-Guimera A (2003) Population deconcentration and social restructuring in Barcelona, a European Mediterranean city. Cities 20(6):387–394
Duvernoy I, Zambon I, Sateriano A, Salvati L (2018) Pictures from the other side of the fringe: urban growth and peri-urban agriculture in a post-industrial city (Toulouse, France). J Rural Stud 57:25–35
e Silva JDA, Correia M (2023) The main drivers of urban sprawl in Portuguese medium cities between 2001 and 2011. Land Use Policy 132:106803
Economou D (1997) The planning system and rural land use control in Greece: a European perspective. Eur Plan Stud 5(4):461–476
El Menchawy A, Aly SS, Hakim MA (2011) The impact of urban sprawl on the heritage areas through the urban fabric of cities. WIT Trans Ecol Environ 150:299–314
Emmanuel D (2014) The Greek system of home ownership and the post-2008 crisis in Athens. Région Et Développement 39:167–182
Escudero Gómez LA, García González JA, Martínez Navarro JM (2019) Medium-sized cities in Spain and their urban areas within national network. Urban Sci 3(1):5
European Environment Agency (2006) Urban sprawl in Europe—The ignored challenge. EEA Report n, Copenhagen, p 10
Fekade W (2000) Deficits of formal urban land management and informal responses under rapid urban growth, an international perspective. Habitat Int 24(2):127–150
Feleki E, Achillas C, Vlachokostas C, Michailidou AV, Ortega L, Moussiopoulos N (2018) Preservation of the Mediterranean identity: an intra-city analysis towards a macro-regional approach for the characterisation of urban sustainability. Sustainability 10(10):3551
Feng Y, Wang X, Du W, Liu J, Li Y (2019) Spatiotemporal characteristics and driving forces of urban sprawl in China during 2003–2017. J Clean Prod 241:118061
Filandri M, Olagnero M (2014) Housing inequality and social class in Europe. Hous Stud 29(7):977–993
Florida R (2011) The great reset. Harper Collins, New York
Fox DM, Witz E, Blanc V, Soulié C, Penalver-Navarro M, Dervieux A (2012) A case study of land cover change (1950–2003) and runoff in a Mediterranean catchment. Appl Geogr 32(2):810–821

Fuladlu K, Riza M, Ilkan M (2021) Monitoring urban Sprawl using time-series data: Famagusta region of Northern Cyprus. SAGE Open 11(2):21582440211007464
Gabriel SA, Rosenthal SS (2015) The boom, the bust and the future of homeownership. Real Estate Econ 43(2):334–374
Galster G, Hanson R, Ratcliffe MR, Wolman H, Coleman S, Freihage J (2001) Wrestling sprawl to the ground: defining and measuring an elusive concept. Hous Policy Debate 12(4):681–717
Gantar D, Golobic M (2015) Landscape scenarios: a study of influences on attitudes and actions in a rural landscape. Futures 69:1–13
Garcia M (2010) The breakdown of the Spanish urban growth model: social and territorial effects of the global crisis. Int J Urban Reg Res 34(4):967–980
Garcia D, Riera P (2003) Expansion versus density in Barcelona: a valuation exercise. Urban Studies 40(10):1925–1936
García-Ayllón S (2013) Retrospective analysis of urban development in the Spanish Mediterranean coast. WIT Trans Ecol Environ 179:291–302
Garcia-López MÀ (2019) All roads lead to Rome… and to sprawl? Evidence from European cities. Reg Sci Urban Econ 79:103467
Garcia-López MA, Muñiz I (2010) Employment decentralisation: polycentricity or scatteration? The case of Barcelona. Urban Stud 47(14):3035–3056
García-Nieto AP, Geijzendorffer IR, Baró F, Roche PK, Bondeau A, Cramer W (2018) Impacts of urbanization around Mediterranean cities: changes in ecosystem service supply. Ecol Ind 91:589–606
Garcia-Ramon MD, Albet A (2000) Pre-Olympic and post-Olympic Barcelona, a model for urban regeneration today? Environ Plan A 32:1331–1334
Gavalas VS, Rontos K, Salvati L (2014) Who becomes an unwed mother in Greece? Socio-demographic and geographical aspects of an emerging phenomenon. Popul Space Place 20(3):250–263
Giannakourou G (2005) Transforming spatial planning policy in Mediterranean countries: Europeanization and domestic change. Eur Plan Stud 13(2):319–331
Gielen E, Riutort-Mayol G, Palencia-Jiménez JS, Cantarino I (2018) An urban sprawl index based on multivariate and Bayesian factor analysis with application at the municipality level in Valencia. Environ Plann B Urban Anal City Sci 45(5):888–914
Gil-Alonso F, Bayona-i-Carrasco J, Pujadas-i-Rúbies I (2016) From boom to crash: Spanish urban areas in a decade of change (2001–2011). Eur Urban Reg Stud 23(2):198–216
Giordano B, Roller E (2004) "Té para todos'? A comparison of the processes of devolution in Spain and the UK. Environ Plann A 36(12)
Gkartzios M, Scott K (2015) A cultural panic in the province? Counterurban mobilities, creativity, and crisis in Greece. Popul Space Place 21(8):843–855
Goldblum C, Wong TC (2000) Growth, crisis and spatial change: a study of haphazard urbanisation in Jakarta, Indonesia. Land Use Policy 17(1):29–37
Gonzales S (2011) The north/south divide in Italy and England: discursive construction of regional inequality. Eur Urban Reg Stud 18(1):62–76
Gordon P, Richardson HW (1996) Beyond polycentricity: the dispersed metropolis, Los Angeles, 1970–1990. J Am Plann Assoc 62(3)
Gospodini A (2001) Urban design, urban space morphology, urban tourism: an emerging new paradigm concerning their relationship. Eur Plan Stud 9(7):925–934
Gospodini A (2006) Portraying, classifying and understanding the emerging landscapes in the post-industrial city. Cities 23(5):311–330
Gospodini A (2009) Post-industrial trajectories of Mediterranean European cities: the case of post-Olympics Athens. Urban Stud 46(5–6):1157–1186
Greenberg M (2021) Seeking shelter: how housing and urban exclusion shape exurban disaster. Sociologica 15(1):67–89
Grout C, Cavailhès J, Détang-Dessendre C, Thomas A (2016) Is Sprawling residential behavior influenced by climate? Land Econ 92(2):203–219

Haase A, Kabisch S, Steinführer A, Bouzarovski S, Hall R, Ogden P (2010) Spaces of reurbanisation: exploring the demographic dimension of inner-city residential change in a European setting. Popul Space Place 16(5):443–463
Habibi S, Asadi N (2011) Causes, results and methods of controlling urban sprawl. Proc Eng 21:133–141
Hall P (1997) Modelling the post-industrial city. Futures 29(4–5):311–322
Haslauer E, Biberacher M, Blaschke T (2012) GIS-based backcasting: an innovative method for parameterisation of sustainable spatial planning and resource management. Futures 44:292–302
Henderson JV (1991) Optimal regulation of land development through price and fiscal controls. J Urban Econ 30:64–82
Hennig EI, Schwick C, Soukup T, Orlitová E, Kienast F, Jaeger JA (2015) Multi-scale analysis of urban sprawl in Europe: towards a European de-sprawling strategy. Land Use Policy 49:483–498
Herold M, Scepan J, Clarke KC (2002) The use of remote sensing and urban landscape metrics to describe structures and changes in urban land uses. Environ Plan A 34(8):1443–1458
Herold M, Goldstein NC, Clarke KC (2003) The spatiotemporal form of urban growth: measurement, analysis and modeling. Remote Sens Environ 86(3):286–302
Herold MH, Couclelis C, Clarke KC (2005) The role of spatial metrics in the analysis and modeling of urban land use change. Comput Environ Urban Syst 29(4):369–399
Hof A, Blázquez-Salom M (2013) The linkages between real estate tourism and urban sprawl in Majorca (Balearic Islands, Spain). Land 2(2):252–277
Inglis NC, Vukomanovic J, Petrasova A, Meentemeyer RK (2023) Viewscape change highlights shifting drivers of exurban development over time. Landsc Urban Plan 238:104833
Ismael HM (2021) Urban form study: the sprawling city—review of methods of studying urban sprawl. GeoJournal 86(4):1785–1796
Johnson MP (2001) Environmental impacts of urban sprawl: a survey of the literature and proposed research agenda. Environ Plan A 33(4):717–735
Kabisch N, Haase D (2011) Diversifying European agglomerations: evidence of urban population trends for the 21st century. Popul Space Place 17(3):236–253
Kaika M (2012) The economic crisis seen from the everyday: Europe's nouveau poor and the global affective implications of a 'local' debt crisis. City 16(4):422–430
Kandylis G, Maloutas T, Sayas J (2012) Immigration, inequality and diversity: socio-ethnic hierarchy and spatial organization in Athens, Greece. Eur Urban Reg Stud 19(3):267–286
Kaniewski D, Van Campo E, Morhange C, Guiot J, Zviely D, Shaked I, Artzy M (2013) Early urban impact on Mediterranean coastal environments. Sci Rep 3(1):3540
Karadimitriou N, Maloutas T, Arapoglou VP (2021) Multiple deprivation and urban development in Athens, Greece: spatial trends and the role of access to housing. Land 10(3):290
Katsibokis G (2013) Ktiriothiki: the architectural heritage of Athens, 1830–1950. J Mod Greek Stud 31(1):133–149
Kazemzadeh-Zow A, Zanganeh Shahraki S, Salvati L, Samani NN (2017) A spatial zoning approach to calibrate and validate urban growth models. Int J Geogr Inf Sci 31(4):763–782
Keddie J, Tonkiss F (2010) The market and the plan: housing, urban renewal and socio-economic change in London. City Cult Soc 1(2):57–67
Kemeny T, Storper M (2012) The sources of urban development: wage, housing, and amenity gaps across American cities. J Reg Sci 52(1):85–108
Kloosterman RC, Musterd S (2001) The polycentric urban region: towards a research agenda. Urban Stud 38(4):623–633
Knieling J (2015) Cities in crisis: socio-spatial impacts of the economic crisis in Southern European cities. Routledge, London
Koomen E (2007) Modelling land-use change: progress and applications. Springer
Koprowska K, Łaszkiewicz E, Kronenberg J (2020) Is urban sprawl linked to green space availability? Ecol Ind 108:105723
Kordel S, Naumann M (2023) The rural housing crisis: analytical dimensions and emblematic issues. Hous Stud 1–18

Kourliouros E (1997) Planning industrial location in Greater Athens: the interaction between deindustrialization and anti-industrialism during the 1980s. Eur Plan Stud 5(4):435–460

Kourliouros E (2003) Reflections on the economic-noneconomic debate: a radical geographical perspective from the European south. Antipode 35(4):781–799

Kovács Z, Farkas ZJ, Egedy T, Kondor AC, Szabó B, Lennert J, Kohán B (2019) Urban sprawl and land conversion in post-socialist cities: the case of metropolitan Budapest. Cities 92:71–81

Kroll F, Kabisch N (2012) The relation of diverging urban growth processes and demographic change along an urban-rural gradient. Popul Space Place 18:260–276

Lagarias A (2023) Impervious land expansion as a control parameter for climate-resilient planning on the Mediterranean Coast: evidence from Greece. Land 12(10):1844

Lagarias A, Prastacos P (2020) Comparing the urban form of South European cities using fractal dimensions. Environ Plann B Urban Anal City Sci 47(7):1149–1166

Lagarias A, Sayas J (2018) Urban sprawl in the Mediterranean: evidence from coastal medium-sized cities. Reg Sci Inquiry 10(3):15–32

Lagarias A, Sayas J (2019) Is there a common typology of urban sprawl in Mediterranean cities? A comparative analysis utilizing spatial metrics. Revue D'économie Régionale & Urbaine 4:813–850

Lagarias A, Stratigea A (2023) Coastalization patterns in the Mediterranean: a spatiotemporal analysis of coastal urban sprawl in tourism destination areas. GeoJournal 88(3):2529–2552

Le Goix R (2005) Gated communities: sprawl and social segregation in Southern California. Hous Stud 20(2):323–343

Lee S (2011) Metropolitan growth patterns and socio-economic disparity in six US metropolitan areas 1970–2000. Int J Urban Reg Res 35(5):988–1011

Leontidou L (1996) Alternatives to modernism in (southern) urban theory: exploring in-between spaces. Int J Urban Reg Res 20(2):178–195

Leontidou L (2009) Beyond the borders of Mediterranean cities: the Mediterranean city in transition. Quart Int Sociol 18(3–4):131–140

Lindstrom MJ, Bartling H (2003) Suburban sprawl: culture, theory, and politics. Rowman & Littlefield, New York

Liu L, Meng L (2020) Patterns of urban sprawl from a global perspective. J Urban Plan Dev 146(2):04020004

Longhi C, Musolesi A (2007) European cities in the process of economic integration: towards structural convergence. Ann Reg Sci 41(2):333–351

Luger J, Schwarze T (2021) Cafés, cocktail coves, and "empathy walls": comparing urban and exurban everyday life through a Lefebvrian lens. Geoforum 127:151–161

MacKinnon D, Cumbers A, Pike A, Birch K, McMaster R (2009) Evolution in economic geography: institutions, political economy and adaptation. Econ Geogr 85:129–150

Malheiros J (2002) Ethni-cities: residential patterns in the Northern European and Mediterranean Metropolises—implications for policy design. Int J Popul Geogr 8(2):107–134

Maloutas T (2003) Promoting social sustainability. The case of Athens. City 7(2):167–181

Maloutas T (2007) Socio-economic classification models and contextual difference: the "European socio-economic classes" (ESeC) from a South European angle. South Eur Soc Polit 12(4):443–460

Maloutas T (2004). Segregation and residential mobility. Socially entrapped social mobility and its impact on segregation in Athens. Eur Urban Reg Stud 11(2).171–187

Marmaras E (1989) The privately-built multi-storey apartment building: the case of inter-war Athens. Plann Perspect 4(1):45–78

Marraccini E, Debolini M, Moulery M, Abrantes P, Bouchier A, Chéry JP, Napoleone C (2015) Common features and different trajectories of land cover changes in six Western Mediterranean urban regions. Appl Geogr 62:347–356

Marshall T (1995) Regional planning in Catalonia. Eur Plan Stud 3:25–45

Martí P, Nolasco-Cirugeda A, Serrano-Estrada L (2017) Assessment tools for urban sustainability policies in Spanish Mediterranean tourist areas. Land Use Policy 67:625–639

Masini E, Tomao A, Barbati A, Corona P, Serra P, Salvati L (2019) Urban growth, land-use efficiency and local socioeconomic context: a comparative analysis of 417 metropolitan regions in Europe. Environ Manage 63:322–337
Masoumi HE (2014) Urban sprawl in mid-sized cities of MENA, evidence from Yazd and Kashan in central Iran. Manage Res Pract 6(2):25–41
Mayer CJ, Somerville CT (2000) Land use regulation and new construction. Reg Sci Urban Econ 30(6):639–662
McCarthy J (2012) The financial crisis and environmental governance 'after' neoliberalism. Tijdschr Econ Soc Geogr 103(2):180–195
McMillen DP, McDonald JF (1993) Could zoning have increased land values in Chicago? J Urban Econ 33:167–188
Mehriar M, Masoumi H, Mohino I (2020) Urban sprawl, socioeconomic features, and travel patterns in Middle-East countries: a case study in Iran. Sustainability 12(22):9620
Mills ES (1967) An aggregative model of resource allocation in a metropolitan area. Am Econ Rev 57:197–210
Monclus F (2000) Barcelona's planning strategies: from 'Paris of the south to capital of the west Mediterranean. GeoJournal 51:57–63
Monclus FJ (2003) The Barcelona model: an original formula? From 'reconstruction' to strategic urban projects (1979–2004). Plan Perspect 18(4):399–421
Morelli VG, Rontos K, Salvati L (2014) Between suburbanisation and re-urbanisation: revisiting the urban life cycle in a Mediterranean compact city. Urban Res Pract 7(1):74–88
Morote ÁF, Hernández M (2016) Urban sprawl and its effects on water demand: a case study of Alicante, Spain. Land Use Policy 50:352–362
Mouratidis K (2019) Compact city, urban sprawl, and subjective well-being. Cities 92:261–272
Mudu P (2006) Patterns of segregation in contemporary Rome. Urban Geogr 27(5):422–440
Muñiz I, Garcia-Lopez MA (2010) The polycentric knowledge economy in Barcelona. Urban Geogr 31(6):774–799
Muñiz I, Galindo A, Garcia-Lopez MA (2003) Cubic spline population density functions and satellite cities delimitation: the case of Barcelona. Urban Stud 40(7):1303–1321
Muñoz F (2003) Lock living: urban sprawl in Mediterranean cities. Cities 20(6):381–385
Muñoz F (2010) Urbanalization: common landscapes, global places. Open Urban Stud J 3:78–88
Musterd S (2005) Social and ethnic segregation in europe: levels, causes and effects. J Urban Aff 27(3):331–348
Musterd S, Marcińczak S, Van Ham M, Tammaru T (2017) Socioeconomic segregation in European capital cities. Increasing separation between poor and rich. Urban Geogr 38(7):1062–1083
Nel-lo O (2011) The five challenges of urban rehabilitation. The Catalan experience. Urban Res Pract 4(3):308–325
Nelson AC, Burby RJ, Feser E, Dawkins CJ, Malizia EE, Quercia R (2004) Urban containment and central-city revitalization. J Am Plann Assoc 70(4):411–425
Newman P, Kenworthy J (1999) Sustainability and cities: overcoming automobile, dependence. Island Press Books, Washington, DC
Newman ME, McLaren KP, Wilson BS (2014) Assessing deforestation and fragmentation in a tropical moist forest over 68 years; the impact of roads and legal protection in the Cockpit country, Jamaica. For Ecol Manage 315:138–152
Nickayin SS, Tomao A, Quaranta G, Salvati L, Gimenez Morera A (2020) Going toward resilience? Town planning, peri-urban landscapes, and the expansion of Athens, Greece. Sustainability 12(24):10471
Nilsson K, Pauleit S, Bell S, Aalbers C, Sick Nielsen TA (2013) Peri-urban futures: scenarios and models for land use change in Europe. Springer-Verlag, Heidelberg, Berlin. https://doi.org/10.1007/978-3-642-30529-0
Orenstein DE, Frenkel A, Jahshan F (2014) Methodology matters: measuring urban spatial development using alternative methods. Environ Plann B Plann des 41:3–23

Oueslati W, Alvanides S, Garrod G (2015) Determinants of urban sprawl in European cities. Urban Stud 52(9):1594–1614

Overmars KP, Verburg PH, Veldkamp A (2007) Comparison of a deductive and an inductive approach to specify land suitability in a spatially explicit land use model. Land Use Policy 24:584–599

Pace A, Attard M, Camilleri M, Valentino G (2023) Urban growth in a Mediterranean Island-State: a data-driven study of Malta's development permits in the last thirty years. Sustainability 15(22):16063

Pacione M (2003) Urban environmental quality and human wellbeing—a social geographical perspective. Landsc Urban Plan 65:19–30

Pagliarin S (2018) Linking processes and patterns: spatial planning, governance and urban sprawl in the Barcelona and Milan metropolitan regions. Urban Stud 55(16):3650–3668

Palang H, Peil T (2010) Mapping future through the study of the past and present: Estonian suburbia. Futures 42:700–710

Panori A, Psycharis Y, Ballas D (2019) Spatial segregation and migration in the city of Athens: investigating the evolution of urban socio-spatial immigrant structures. Popul Space Place 25(5):e2209

Parcerisas L, Marull J, Pino J, Tello E, Coll F, Basnou C (2012) Land use changes, landscape ecology and their socioeconomic driving forces in the Spanish Mediterranean coast (El Maresme County, 1850–2005). Environ Sci Policy 23:120–132

Pardo-García S, Mérida-Rodríguez M (2018) Physical location factors of metropolitan and rural sprawl: geostatistical analysis of three Mediterranean areas in Southern Spain. Cities 79:178–186

Parés M, March H, Saurí D (2013) Atlantic gardens in Mediterranean climates: understanding the production of suburban natures in Barcelona. Int J Urban Reg Res 37:328–347

Paül V, Tonts M (2005) Containing urban sprawl: trends in land use and spatial planning in the metropolitan region of Barcelona. J Environ Planning Manage 48(1):7–35

Pérez JMG (2010) The real estate and economic crisis: an opportunity for urban return and rehabilitation policies in Spain. Sustainability 2(6):1571–1601

Perrin C, Nougarèdes B, Sini L, Branduini P, Salvati L (2018) Governance changes in peri-urban farmland protection following decentralisation: a comparison between Montpellier (France) and Rome (Italy). Land Use Policy 70:535–546

Petrov L, Lavalle C, Kasanko M (2009) Urban land use scenarios for a tourist region in Europe: applying the MOLAND model to Algarve, Portugal. Landsc Urban Plann 92:10–23

Petsimeris P, Tsoulouvis L (1997) Current trends and prospects of future change in the Greek urban system. Rivista Geografica Italiana 104(4):421–443

Petsimeris P (2002) Population deconcentration in Italy, Spain and Greece: a first comparison. Ekistics 163–172

Phillips J, Goodstein E (2000) Growth management and housing prices: the case of Portland, Oregon. Contemp Econ Policy 18:334–344

Polet R, Nomden K (1997) Improving the management of European regional development policy. Eur Plan Stud 5:777–792

Polyzos S, Minetos D, Niavis S (2013) Driving factors and empirical analysis of urban sprawl in Greece. Theor Empir Res Urban Manage 8(1):5–29

Ponstingel D (2020) The impact of exurban development on forested areas in Kurgan City, Russia. Land Use Policy 94:104485

Power A (2001) Social exclusion and urban sprawl: is the rescue of cities possible? Reg Stud 35(8):731–742

Pozoukidou G, Ntriankos I (2017) Measuring and assessing urban sprawl: a proposed indicator system for the city of Thessaloniki, Greece. Remote Sens Appl Soc Environ 8:30–40

Puglisi M, Marvin S (2002) Developing urban and regional foresight: exploring capacities and identifying needs in the North West. Futures 34:761–777

Rappert B (1999) Rationalising the future? Foresight in science and technology policy co-ordination. Futures 31:527–545

Ratcliffe J, Krawczyk E (2011) Imagineering city futures: the use of prospective through scenarios in urban planning. Futures 43:642–653

Recanatesi F, Clemente M, Grigoriadis E, Ranalli F, Zitti M, Salvati L (2016) A fifty-year sustainability assessment of Italian agro-forest districts. Sustainability 8(1):32

Rejeb Bouzgarrou A, Claramunt C, Rejeb H (2019) Visualizing urban sprawl effects of a Tunisian city: a new urban spatial configuration of Monastir. Ann GIS 25(1):71–82

Remoundou K, Gkartzios M, Garrod G (2016) Conceptualizing mobility in times of crisis: towards crisis-led counterurbanization? Reg Stud 50(10):1663–1674

Ren X (2015) City power and urban fiscal crises: the USA, China, and India. Int J Urban Sci 19(1):73–81

Richardson H, Chang-Hee C (2004) Urban sprawl in Western Europe and the United States. Ashgate, London

Rienow A, Goetzke R (2015) Supporting SLEUTH—enhancing a cellular automaton with support vector machines for urban growth modeling. Comput Environ Urban Syst 49:66–81

Rijkens-Klomp N, Van Der Duin P (2014) Evaluating local and national public foresight studies from a user perspective. Futures 59:18–26

Robert S, Fox D, Boulay G, Grandclement A, Garrido M, Pasqualini V, Trémélo ML (2019) A framework to analyse urban sprawl in the French Mediterranean coastal zone. Reg Environ Change 19:559–572

Rontos K, Grigoriadis S, Sateriano A, Syrmali M, Vavouras I, Salvati L (2016) Lost in Protest, Found in Segregation: Divided Cities in the Light of the 2015 'Oki' Referendum in Greece. City Cult Soc 7(3):139–148

Roubien D (2020) Creating modern Athens: a capital between east and west. Routledge, London

Rounsevell MDA, Ewert F, Reginster I, Leemans R, Carter TR (2005) Future scenarios of European agricultural land use: II. Projecting changes in cropland and grassland. Agr Ecosyst Environ 107:117–135

Rubiera Morollón F, González Marroquin VM, Pérez Rivero JL (2016) Urban sprawl in Spain: differences among cities and causes. Eur Plan Stud 24(1):207–226

Rubiera-Morollón F, Garrido-Yserte R (2020) Recent literature about urban sprawl: A renewed relevance of the phenomenon from the perspective of environmental sustainability. Sustainability 12(16):6551

Saizen I, Mizuno K, Kobayashi S (2006) Effects of land-use master plans in the metropolitan fringe of Japan. Landsc Urban Plan 78:411–421

Sakieh Y, Salmanmahiny A, Jafarnezhad J, Mehri A, Kamyab H, Galdavi S (2015) Evaluating the strategy of decentralized urban land-use planning in a developing region. Land Use Policy 48:534–551

Salem M, Tsurusaki N, Divigalpitiya P (2020) Land use/land cover change detection and urban sprawl in the peri-urban area of greater Cairo since the Egyptian revolution of 2011. J Land Use Sci 15(5):592–606

Salvati L (2013) Long-term dynamics of self-contained urban growth in compact and dispersed regions of southern Europe. Land Use Policy 35:213–225

Salvati L (2014) Towards a polycentric region? The socioeconomic trajectory of Rome, an 'eternal' Mediterranean City. Tijdschr Econ Soc Geogr 105(3):268–284

Salvati L (2016) The dark side of the crisis: Disparities in per Capita income (2000–12) and the Urban-Rural Gradient in Greece. Tijdschr Econ Soc Geogr 107(5):628–641

Salvati L (2019) Urban dispersion and economic crisis: empirical evidence from a Mediterranean region. J Environ Planning Manage 62(7):1205–1226

Salvati L, Carlucci M (2014) Distance matters: land consumption and the mono-centric model in two Southern European cities. Landsc Urban Plan 27:41–51

Salvati L, Carlucci M (2016) Patterns of sprawl: the socioeconomic and territorial profile of dispersed urban areas in Italy. Reg Stud 50(8):1346–1359

Salvati L, De Rosa S (2014) 'Hidden polycentrism' or 'subtle dispersion'? Urban growth and long-term sub-centre dynamics in three Mediterranean cities. Land Use Policy 39:233–243

Salvati L, Gargiulo Morelli V (2014) Unveiling urban sprawl in the Mediterranean region: towards a latent urban transformation? Int J Urban Reg Res 38(6):1935–1953

Salvati L, Zitti M (2009) The environmental "risky" region: identifying land degradation processes through integration of socio-economic and ecological indicators in a multivariate regionalization model. Environ Manage 44(5):888

Salvati L, Gemmiti R, Perini L (2012) Land degradation in Mediterranean urban areas: an unexplored link with planning? Area 44(3):317–325

Salvati L, Zitti M, Sateriano A (2013) Changes in city vertical profile as an indicator of sprawl: evidence from a Mediterranean urban region. Habitat Int 38:119–125

Salvati L, Sateriano A, Grigoriadis E (2016) Crisis and the city: profiling urban growth under economic expansion and stagnation. Lett Spat Resour Sci 9(3):329–342

Salvati L, Ferrara A, Chelli F (2018) Long-term growth and metropolitan spatial structures: an analysis of factors influencing urban patch size under different economic cycles. Geografisk Tidsskrift—Danish J Geogr 118(1):56–71

Salvati L, Ciommi MT, Serra P, Chelli FM (2019) Exploring the spatial structure of housing prices under economic expansion and stagnation: the role of socio-demographic factors in metropolitan Rome, Italy. Land Use Policy 81:143–152

Samet RH (2013) Complexity, the science of cities and long-range futures. Futures 47:49–58

Sampson RJ, Schachner JN, Mare RD (2017) Urban income inequality and the great recession in Sunbelt form: disentangling individual and neighborhood-level change in Los Angeles. Russell Sage Found J Social Sci 3(2):102–128

Sanz ASR, Fernandez C, Mouillot F, Ferrat L, Istria D, Pasqualini V (2013) Long-term forest dynamics and land-use abandonment in the Mediterranean Mountains, Corsica, France. Ecol Soc 18(2) (art. no. 38). https://doi.org/10.5751/ES-05556-180238

Sardà R, Avila C, Mora J (2005) A methodological approach to be used in integrated coastal zone management processes: The case of the Catalan Coast (Catalonia, Spain). Estuar Coast Shelf Sci 62(3):427–439

Sassen S (2010) The city: Its return as a lens for social theory. City Cult Soc 1(1):3–11

Saurí D (2003) Lights and shadows of urban water demand management. The case of the metropolitan region of Barcelona. Eur Plann Stud 11(3):229–243

Schneider F, Kallis G, Martinez-Alier J (2010) Crisis or opportunity? Economic degrowth for social equity and ecological sustainability. Introduction to this special issue. J Clean Prod 18(6):511–518

Schneider A, Woodcock CE (2008) Compact, dispersed, fragmented, extensive? A comparison of urban growth in twenty-five global cities using remotely sensed data, pattern metrics and census information. Urban Stud 45(3):659–692

Schwaab J (2022) Sprawl or compactness? How urban form influences urban surface temperatures in Europe. City Environ Interact 16:100091

Scott AJ, Storper M (2015) The nature of cities: the scope and limits of urban theory. Int J Urban Reg Res 39(1):1–15

Serra P, Pons X, Saurí D (2008) Land-cover and land-use change in a Mediterranean landscape: a spatial analysis of driving forces integrating biophysical and human factors. Appl Geogr 28:189–209

Serra P, Vera A, Tulla AF (2013) Spatial and socio-environmental dynamics of Catalan regional planning from a multivariate statistical analysis using 1980s and 2000s data. Eur Plan Stud 22(6):1280–1300

Serra P, Vera A, Tulla AF, Salvati L (2014) Beyond urban-rural dichotomy: exploring socioeconomic and land-use processes of change in Spain (1991–2011). Appl Geogr 55:71–81

Sharaf SA, Serra P, Saurí D (2018) A district and sector land-use and landscape analysis of urban sprawl in Al Ain municipality (United Arab Emirates): just a quick conversion from sand to a built-up environment? Appl Geogr 95:88–100

Sharma T, Carmichael J, Klinkenberg B (2006) Integrated modeling for exploring sustainable agriculture futures. Futures 38:93–113

Shearer AW (2005) Approaching scenario based studies: three perceptions about the future and considerations for landscape planning. Environ Plann B Plann des 32(1):67–87
Sicular T, Ximing Y, Gustafsson B, Shi L (2007) The urban-rural income gap and inequality in China. Rev Income Wealth 53(1):93–126
Silva EA (2004) The DNA of our regions: artificial intelligence in regional planning. Futures 36:1077–1094
Silva EA, Clarke KC (2002) Calibration of the SLEUTH urban growth model for Lisbon and Porto, Portugal. Comput Environ Urban Syst 26:525–552
Skayannis P (2013) The (master) plans of Athens and the challenges of its re-planning in the context of crisis. ArchNet-IJAR Int J Arch Res 7(2):192
Slaughter R (2002) Futures studies as a civilisational catalyst. Futures 34(3–4):349–353
Smiraglia D, Ceccarelli T, Bajocco S, Salvati L, Perini L (2016) Linking trajectories of land change, land degradation processes and ecosystem services. Environ Res 147:590–600
Smith A (2005) Conceptualizing city image change: the 're-imaging' of Barcelona. Tour Geogr 7(4):398–423
Solecki WD, Oliveri C (2004) Downscaling climate change scenarios in an urban land use change model. J Environ Manage 72:105–115
Song Y, Knaap G-J (2004) Measuring urban form: Is Portland winning the war on sprawl? J Am Plann Assoc 70(2):210–225
Soulard CT, Valette E, Perrin C, Abrantes PC, Anthopoulou T, Benjaballah O, Paoli JC (2018) Peri-urban agro-ecosystems in the Mediterranean: diversity, dynamics, and drivers. Reg Environ Change 18:651–662
Souliotis N (2013) Cultural economy, sovereign debt crisis and the importance of local contexts: the case of Athens. Cities 33:61–68
Stathakis D, Baltas P (2024) The Greek model of urbanization. Land Use Policy 140:107113
Sterk B, van Ittersum MK, Leeuwis C (2011) How, when, and for what reasons does land use modelling contribute to societal problem solving? Environ Model Softw 26:310–316
Stewart, (2008) Integral scenarios: reframing theory, building from the practice. Futures 40:160–172
Storper M, Scott AJ (2016) Current debate in urban theory: a critical assessment. Urban Stud 53(6):1114–1136
Swyngedouw E (2004) Social power and the urbanization of water. Oxford University Press, Oxford
Tomao A, Quaranta G, Salvia R, Vinci S, Salvati L (2021) Revisiting the 'southern mood'? Post-crisis Mediterranean urbanities between economic downturns and land-use change. Land Use Policy 111:105740
Townsend A (2009) Integration of economic and spatial planning across scales. Int J Public Sect Manag 22(7):643–659
Triantakonstantis D, Stathakis D (2015) Examining urban sprawl in Europe using spatial metrics. Geocarto Int 30(10):1092–1112
Tsai Y (2005) Quantifying urban form: compactness versus sprawl. Urban Stud 42(1):141–161
Tsapala F, Chorianopoulos I, Gaki E, Gialis S (2023) The Mediterranean City and EU integration: divergent resurgence paths and new forms of north-south urban divide. Cities 138:104352
Tsiligiannis A (2020) Why isn't urban development sustainable? An institutional approach to the case of Athens, Greece. J Contemp Urban Affairs 4(1):71–78
Tsilimigkas G, Kizos T (2014) Space, pressures and the management of the Greek landscape. Geografiska Annaler Ser B Human Geogr 96(2):159–175
Tsilimigkas G, Stathakis D, Pafi M (2016) Evaluating the land use patterns of medium-sized Hellenic cities. Urban Res Pract 9(2):181–203
Tsilimigkas G, Kizos T, Gourgiotis A (2018) Unregulated urban sprawl and spatial distribution of fire events: evidence from Greece. Environ Hazards 17(5):436–455
Tsilimigkas G, Gourgiotis A, Derdemezi ET (2022) Spatial planning incompetence to discourage urban sprawl on Greek Islands. Evidence from Paros, Greece. J Coastal Conserv 26(2):11
Turok I, Mykhnenko V (2007) The trajectories of European cities, 1960–2005. Cities 24:165–182

Uotila T, Melkas H, Harmaakorpi V (2005) Incorporating futures research into regional knowledge creation and management. Futures 37(8):849–866

Vaiou D (1997) Facets of spatial development and planning in Greece. Eur Plan Stud 5:431–433

Vaiou D (2002) Milestones in the urban history of Athens. Treballs de la Societat Catalana de Geografia 209–226

van Criekingen M (2010) Gentrifying the re-urbanisation debate, not vice versa: the uneven socio-spatial implications of changing transitions to adulthood in Brussels. Popul Space Place 16(5):381–394

van Vliet J, Naus N, van Lammeren RJA, Bregt AK, Hurkens J, van Delden H (2013) Measuring the neighbourhood effect to calibrate land use models. Comput Environ Urban Syst 41(55):64

van Vliet J, Bregt AK, Brown DG, van Delden H, Heckbert S, Verburg PH (2016) A review of current calibration and validation practices in land-change modeling. Environ Model Softw 82:174–182

Varela-Candamio L, Rubiera Morollón F, Sedrakyan G (2019) Urban sprawl and local fiscal burden: analysing the Spanish case. Empirica 46(1):177–203

Vaz E, Painho M, Nijkamp P (2015) Linking agricultural policies with decision-making: a spatial approach. Eur Plan Stud 23(4):733–745

Veldkamp A, Verburg PH (2004) Modelling land use change and environmental impact. J Environ Manage 72(1):1–3

Venanzoni G, Carlucci M, Salvati L (2017) Latent sprawl patterns and the spatial distribution of businesses in a southern European city. Cities 62:50–61

Verburg PH, Schot PP, Dijst MJ, Veldkamp A (2004) Land use change modelling: current practice and research priorities. GeoJournal 61(4):309–324

Vicino TJ, Voigt RH, Kabir M, Michanie J (2022) Urban crises and the Covid-19 pandemic: an analytical framework for metropolitan resiliency. Urban Plann 7(3):4–14

Vidal M, Domene E, Saurí D (2011) Changing geographies of water-related consumption: residential swimming pools in suburban Barcelona. Area 43(1):67–75

Villa P (2012) Mapping urban growth using Soil and Vegetation Index and Landsat data: the Milan (Italy) city area case study. Landsc Urban Plan 107:245–254

Vinci S, Egidi G, López Gay A, Salvati L (2022) Population growth and urban management in metropolitan regions: the contribution of natural balance and migration to polycentric development in Barcelona. Appl Spat Anal Policy 15(1):71–94

Vinci S, Vardopoulos I, Salvati L (2023) A tale of a shrinking City? Exploring the complex interplay of socio-demographic dynamics in the recent development of Attica, Greece. Cities 132:104089

Vradis A (2014) Crisis-scapes suspended: introduction. City 18(4–5):498–501

Weber R (2010) Selling city futures: the financialization of urban redevelopment policy. Econ Geogr 86(3):251–274

Weng YC (2007) Spatiotemporal changes of landscape pattern in response to urbanization. Landsc Urban Plan 81:341–353

Westerink J, Haase D, Bauer A, Ravetz J, Jarrige F, Aalbers CBEM (2013) Dealing with sustainability trade-offs of the compact city in peri-urban planning across European city regions. Eur Plan Stud 21(4):473–497

Wey WM, Hsu J (2014) New urbanism and smart growth: toward achieving a smart National Taipei University District. Habitat Int 42:164–174

Wiek A, Binder C, Scholz RW (2006) Functions of scenarios in transition processes. Futures 38:740–766

Winarso H, Firman T (2002) Residential land development in Jabotabek, Indonesia: triggering economic crisis? Habitat Int 26(4):487–506

Wolff C, Nikoletopoulos T, Hinkel J, Vafeidis AT (2020) Future urban development exacerbates coastal exposure in the Mediterranean. Sci Rep 10(1):14420

Wynn M (1984) Planning and urban growth in Southern Europe. Mansell, London

Xiang W, Clarke KC (2003) The use of scenarios in land-use planning. Environ Plann B Plann des 30:885–909

Yılmaz M, Terzi F (2021) Measuring the patterns of urban spatial growth of coastal cities in developing countries by geospatial metrics. Land Use Policy 107:105487

You T, Zi H (2007) The economic crisis and efficiency change: evidence from the Korean construction industry. Appl Econ 39(14):1833–1842

Zambon I, Salvati L (2019) Metropolitan growth, urban cycles and housing in a Mediterranean country, 1910s–2010s. Cities 95:102412

Zambon I, Serra P, Sauri D, Carlucci M, Salvati L (2017) Beyond the 'Mediterranean city': socioeconomic disparities and urban sprawl in three Southern European cities. Geografiska Annaler Ser B Human Geogr 99(3):319–337

Zambon I, Benedetti A, Ferrara C, Salvati L (2018a) Soil matters? A multivariate analysis of socioeconomic constraints to urban expansion in Mediterranean Europe. Ecol Econ 146:173–183

Zambon I, Serra P, Salvia R, Salvati L (2018b) Fallow land, recession and socio-demographic local contexts: recent dynamics in a Mediterranean urban fringe. Agriculture 8(10):159

Zambon I, Colantoni A, Salvati L (2019) Horizontal vs vertical growth: understanding latent patterns of urban expansion in large metropolitan regions. Sci Total Environ 654:778–785

Zwiers M, Bolt G, van Ham M, Van Kempen R (2016) The global financial crisis and neighbourhood decline. Urban Geogr 37(5):664–684

Index

L. Salvati, *Economic Growth, Regional Disparities, and Urban Sprawl*, Springer Geography, https://doi.org/10.1007/978-3-031-85799-7